Susan S. Adler   Dominiek Beckers
Math Buck

PNF
in Practice

An Illustrated Guide

With 177 Figures
in 464 Separate Illustrations

Springer-Verlag
Berlin Heidelberg New York
London Paris Tokyo
Hong Kong Barcelona
Budapest

Susan S. Adler
4386 Green Valley Rd., Suisun
CA 94585 USA

Dominiek Beckers
Math Buck
Samenwerkende Revalidatiecentra Limburg
Postbus 88
NL-6430 AB Hoensbroek

ISBN 3-540-52 649-8 Springer-Verlag Berlin Heidelberg New York
ISBN 0-387-52 649-8 Springer-Verlag New York Berlin Heidelberg

Library of Congress Cataloging-in-Publication Data
Adler, Susan S., 1931–   . PNF in practice : an illustrated guide / Susan S. Adler, Do-
miniek Beckers, Math Buck. p.   cm.
ISBN 3-540-52649-8 (Berlin : acid-free paper). –– ISBN 0-387-52649-8 (New York : acid-
free paper)
1. Physical therapy. 2. Muscles. 3. Proprioceptors. I. Beckers, Dominiek, 1952–
II. Buck, Math, 1948–   . III. Title.
RM700.A644   1993   615.8'2--dc20   93-5234

The use of general descriptive names, registered names, trademarks, etc. in this publication
does not imply, even in the absence of a specific statement, that such names are exempt
from the relevant protective laws and regulations and therefore free for general use.

Product liability: The publishers cannot guarantee the accuracy of any information about
dosage and application contained in this book. In every individual case the user must check
such information by consulting the relevant literature.

Cover: H. Lopka, Ilvesheim
Typesetting and printing: Appl, Wemding
Binding: Schäffer, Grünstadt
21/3111 – 5 4   – Printed on acid-free paper

To Maggie Knott, teacher and friend.

Devoted to her patients,
dedicated to her students,
a pioneer in her profession

Preface

Proprioceptive neuromuscular facilitation (PNF) is a method of treatment started by Dr. Herman Kabat in the 1940s. Dr. Kabat and Margaret (Maggie) Knott continued to expand and develop the treatment techniques and procedures after their move to Vallejo, California in 1947. After Dorothy Voss joined the team in 1953, Maggie and Dorothy wrote the first PNF book, published in 1956. Dr. Sedgewick Mead, who replaced Dr. Kabat, supported the continued growth of the PNF concept. PNF was originally used as treatment for patients with poliomyelitis. With experience it became clear that this treatment approach was effective in for patients with a wide range of diagnoses.

The three- and six-month PNF courses in Vallejo began in the 1950s. Physical therapists from all over the world have come to Vallejo to learn the theoretical and practical aspects of the PNF concept. In addition, Knott and Voss traveled in the United States and abroad to give introductory courses in the concept.

When Maggie Knott died in 1978 her work at Vallejo was carried on by Carolyn Oei Hvistendahl, who is now living in Norway, and by Hink Mangold, the present director of the PNF program. Sue Adler, Gregg Johnson, and Vicky Saliba continued Maggie's work as teachers of the PNF concept.

Developments in the PNF concept are closely followed in Western Europe, most notably in the United Kingdom and in the Scandinavian and German-speaking countries. It is now possible to take recognized training courses in Europe given by qualified PNF instructors; Bad Ragaz (Switzerland), Berlin and Mainz (Germany), Gothenburg (Sweden), Hoensbroek (the Netherlands) and London (England). In the coming years several other institutes will be qualified to offer PNF courses.

There are other excellent books dealing with the PNF method, but they are all general works providing an extensive theoretical description of the treatment. We felt there was a need for a comprehensive coverage of the practical techniques in text and illustrations. This book should thus be seen as a practical guide and used in combination with existing textbooks.

The aims of this book are:
– To attain a uniformity in practical treatment.
– To record the most recent developments in PNF and put them into word and picture.
– To show the PNF method and, in so doing, help students and practitioners of physical therapy in their PNF training.

The realization of this book would not have been possible without the cooperation of the Lucas Stichting voor Revalidatie in Hoensbroek (the Netherlands). A special note of thanks goes to the following: R. Holthuis and F. Somers for the photography, José van Oppen for acting as a model, Pia van Heel, Jan Albers, and Constance Kusters for the organization, Andrew Davies for translation work and Ben Eisermann for the drawings.

June 1993   S. S. ADLER
                                                       D. BECKERS
                                                       M. BUCK

Contents

X

1 Introduction to Proprioceptive Neuromuscular Facilitation

Proprioceptive: Having to do with any of the sensory receptors that give information concerning movement and position of the body

Neuromuscular: Involving the nerves and muscles

Facilitation: Making easier

Proprioceptive neuromuscular facilitation (PNF) is more than a technique, it is a philosophy of treatment. The basis of this philosophy is the idea that all human beings, including those with disabilities, have untapped existing potential (Kabat 1950). In keeping with this philosophy, there are certain principles that are basic to PNF:

1. The treatment approach is always positive, reinforcing and using that which the patient can do, on a physical and psychological level.
2. The primary goal of all treatment is to help patients achieve their highest level of function.
3. PNF is an integrated approach: each treatment is directed at a total human being, not at a specific problem or body segment.

The material in this book is based on treatment innovations begun by Dr. Herman Kabat and expanded by Margaret Knott, Dorothy Voss, and others, both physical therapists and patients. The authors acknowledge their debt to these outstanding people, and hope that this book will encourage others to carry on the work. We recommend further reading in both the second and third editions of the book *Proprioceptive neuromuscular facilitation: Patterns and Techniques* by Knott and Voss, and Voss, Ionta and Meyers, respectively.

This book covers the procedures, techniques, and patterns within PNF. Their application to patient treatment is discussed throughout with special attention on mat activities, gait, and self-care. The emphasis within this book is twofold: developing an understanding of the principles that underlie PNF, and showing through pictures rather than with words how to perform the patterns and activities. Skill in applying the principles and practices of PNF to patient treatment cannot be learned only from a book. We recommend that the learner combine reading with classroom practice and patient treatment under the supervision of a skilled PNF practitioner.

The work of Sir Charles Sherrington was important in the development of the procedures and techniques of PNF. The following useful definitions were abstracted from his work (Sherrington 1947):

- *Afterdischarge:* The effect of a stimulus continues after the stimulus stops. If the strength and duration of the stimulus increase, the afterdischarge increases also. The feeling of increased power that comes after a maintained static contraction is a result of afterdischarge.
- *Temporal summation:* A succession of weak stimuli (subliminal) occurring within a certain (short) period of time combine (summate) to cause excitation.
- *Spatial summation:* Weak stimuli applied simultaneously to different areas of the body reinforce each other (summate) to cause excitation. Temporal and spatial summation can combine for greater activity.
- *Irradiation:* This is a spreading and increased strength of a response. It occurs when either the number of stimuli or the strength of the stimuli is increased. The response may be either *excitation* or *inhibition.*
- *Successive induction:* An increased excitation of the agonist muscles follows stimulation (contraction) of their antagonists. Techniques involving *reversal of antagonists* make use of this property. (Induction: stimulation, increased excitability.)
- *Reciprocal innervation (reciprocal inhibition):* Contraction of muscles is accompained by simultaneous inhibition of their antagonists. Reciprocal innervation is a necessary part of coordinated motion. *Relaxation techniques* make use of this property.

"The nervous system is continuous throughout its extent – there are no isolated parts."

References

Kabat H (1950) Studies on neuromuscular dysfunktion, XIII: New concepts and techniques of neuromuscular reeducation for paralysis. Perm Found Med Bull 8 (3): 121–143
Sherrington C (1947) The integrative action of the nervous system. Yale University Press, New Haven

2 Basic Procedures for Facilitation

The basic facilitation procedures provide tools for the therapist to help the patient gain efficient motor function. Their effectiveness does *not* depend on having the conscious cooperation of the patient. The procedures are used to:

1. Increase the patient's ability to move or remain stable
2. Guide the motion by proper grips and appropriate resistance
3. Help the patient achieve coordinated motion through timing
4. Increase the patient's stamina and avoid fatigue

The basic procedures overlap in their effects. For example, *resistance* is necessary to make the *stretch reflex* effective (Gellhorn 1949), and the effect of resistance changes with the alignment of the therapist's body and the direction of the Manual contact.

We can use these basic procedures to treat patients with any diagnosis or condition, although a patient's condition may rule out the use of some of them. Basically, the therapist should avoid causing or increasing pain: Pain is an inhibitor of effective and coordinated muscular performance and it can be a sign of potential harm (Hislop 1960; Fisher 1967). Other contraindications are mainly common sense: for example, not using *approximation* on an extremity with an unhealed fracture. In the presence of unstable joints, the therapist should take great care if using the *stretch reflex.*

The basic procedures for facilitation are:

- *Resistance:* To aid muscle contraction and motor control, to increase strength
- *Irradiation and reinforcement:* Use of the spread of the response to stimulation
- *Manual contact:* To increase strength and guide motion with grip and pressure
- *Body position and body mechanics:* Guidance and control of motion by the alignment of the therapist's body, arms, and hands
- *Verbal (commands):* use of words and the appropriate vocal volume to direct the patient
- *Vision:* Use of vision to guide motion and increase force
- *Traction and approximation:* The elongation or compression of the limbs and trunk to facilitate motion and stability
- *Stretch:* The use of muscle elongation and the stretch reflex to facilitate contraction and decrease muscle fatigue
- *Timing:* to Promote normal timing and increase muscle contraction through "timing for emphasis"
- *Patterns:* Synergistic mass movements, components of functional normal motion

2.1 Resistance

Resistance is used in treatment to

1. Facilitate the ability of the muscle to contract
2. Increase motor control
3. Help the patient gain an awareness of motion
4. Increase strength

Most of the PNF techniques evolved from knowing the effects of resistance. Although Kabat, Knott, and Voss used the term *maximal* to describe the proper amount of resistance, many PNF instructors now consider the terms *optimal* or *appropriate* more accurate (G. Johnson and V. Saliba, S. S. Adler, M. L. Mangold, unpublished hand-outs). The amount of resistance provided during an activity must be correct for the patient's, condition and the goal of the activity.

Gellhorn showed that when a muscle contraction is resisted that muscle's response to cortical stimulation increases. The active muscle tension produced by resistance is the most effective proprioceptive facilitation. The magnitude of that facilitation is related directly to the amount of resistance (Gellhorn 1949; Loofbourrow and Gellhorn 1948). The proprioceptive reflexes from contracting muscles increase the response of synergistic muscles[1] at the same joint and associated synergists at neighboring joints. This facilitation can spread from proximal to distal and from distal to proximal. Antagonists of the facilitated muscles are usually inhibited. If the muscle activity in the agonists becomes intense, there may be activity in the antagonistic muscle groups as well (cocontraction). (Gellhorn 1947; Loofbourrow and Gellhorn 1948).

How we give resistance depends on the kind of muscle contraction being resisted (Fig. 2.1). We define the types of muscle contraction as follows (International PNF Instructor Group, unpublished hand-out):

1. *Isotonic (dynamic):* The intent of the patient is to produce motion.
   a) *Concentric:* shortening of the agonist produces motion.
   b) *Eccentric:* an outside force, gravity or resistance, produces the motion. The motion is restrained by the controlled lengthening of the agonist.
   c) *Stabilizing isotonic:* the intent of the patient is motion; the motion is prevented by an outside force (usually resistance).
2. *Isometric (static):* The intent of both the patient and the therapist is that *no* motion occur.

The resistance to concentric or eccentric muscle contractions should be adjusted so that motion can occur in a smooth and coordinated manner. Resistance to a stabilizing contraction must be controlled to maintain the stabilized position. When resisting an isometric contraction, the resistance should be increased and decreased gradually so that no motion occurs. It is important that the resistance does not cause pain or unwanted fatigue. Both the therapist and the patient should avoid breath-holding. Timed and controlled inhalations and exhalations can increase the patient's strength and active range of motion.

---

[1] Synergists are muscles which act with other muscles to produce coordinated motion.

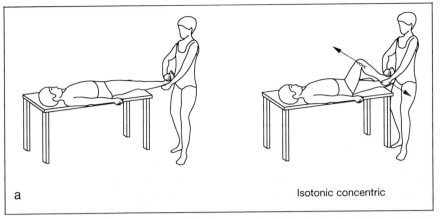

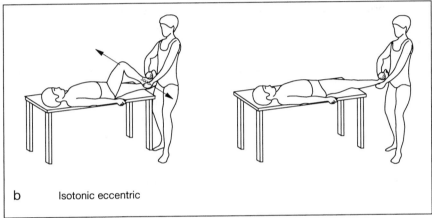

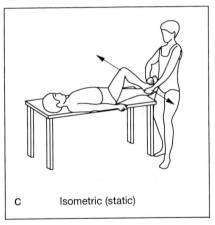

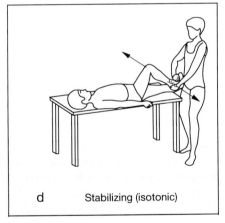

Fig. 2.1 a–d. Types of muscle contraction. *a* Movement into a shortened range. *b* The resistance provided by the physiotherapist is stronger: movement into the lengthened range. *c* The forces exerted by both patient and physiotherapist are the same: there is no intention to move. *d* The patient tries to move but is prevented by the physiotherapist (modified, from Klein-Vogelbach, 1990. Functional Kinetics. Springer Berlin Heidelberg New York)

2.2 Irradiation and Reinforcement

Properly applied resistance results in irradiation and reinforcement. We define *irradiation* as the spread of the response to stimulation. This response can be seen as increased facilitation (contraction) or inhibition (relaxation) in the synergistic muscles and patterns of movement. The response increases as the stimuli increase in intensity or duration (Sherrington 1947). Kabat (1961) wrote that it is resistance to motion that produces irradiation, and the spread of the muscular activity will occur in specific patterns.

Reinforce, as defined in *Webster's Ninth New Collegiate Dictionary,* is "to strengthen by fresh addition, make stronger." The therapist directs the reinforcement of the weaker muscles by the amount of resistance given to the strong muscles.

Increasing the amount of resistance will increase the amount and extent of the muscular response. Changing the movement that is resisted or the position of the patient also will change the results. The therapist adjusts the amount of resistance and type of muscle contraction to suit (1) the condition of the patient, and (2) the goal of the treatment. Because each patient reacts differently, it is not possible to give general instructions on how much resistance to give or which movements to resist. By assessing the results of the treatment, the therapist can determine the best uses of resistance, irradiation, and reinforcement.

Examples of the use of resistance in patient treatment include:

1. Resisting muscle contractions in a sound limb to produce contraction of the muscles in the immobilized contralateral limb.
2. Resisting hip flexion to cause contraction of the trunk flexor muscles (Fig. 2.2).
3. Resisting supination of the forearm so the external rotators of that shoulder will contract.
4. Resisting hip flexion – adduction – external rotation to make the dorsiflexors contract with inversion (Fig. 2.3).

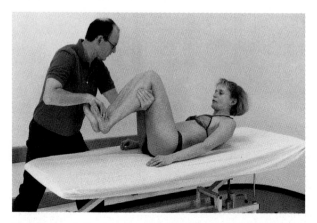

Fig. 2.2. Irradiation into the trunk flexor muscles when doing bilateral leg patterns

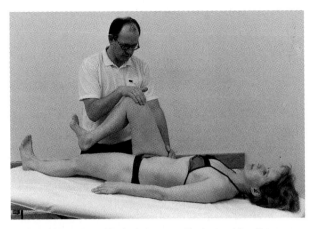

Fig. 2.3. Irradiation to dorsiflexion and inversion with the leg pattern flexion – adduction – external rotation

2.3 Manual Contact

The therapist's grip stimulates the patient's skin receptors and other pressure receptors. The contact should give the patient information about the proper direction of motion. The therapist's hand should be placed to apply the pressure opposite the direction of motion. The sides of the arm or leg are considered neutral surfaces and may be held.

Pressure on a muscle aids that muscle's ability to contract. Putting pressure that is opposite to the direction of motion on any point of the moving limb will stimulate the synergistic muscles to reinforce the movement. Manual contact on the patient's trunk helps the limb motion indirectly by promoting trunk stabilization.

To control movement and resist rotation the therapist uses a *lumbrical grip* (Fig. 2.4). In this grip the pressure comes from flexion at the metacarpal – phalangeal joints, allowing the therapist's fingers to conform to the body part. The lumbrical grip provides the therapist with good control of the motion without causing the patient pain due to squeezing (Fig. 2.5).

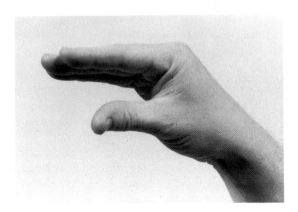

Fig. 2.4. The lumbrical grip

7

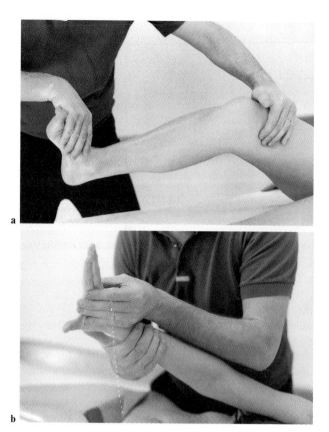

a

b

Fig. 2.5 a, b. Lumbrical grips. *a* For the leg pattern flexion – adduction – external rotation. *b* For the arm pattern flexion – abduction – external rotation

2.4 Body Position and Body Mechanics

Johnson and Saliba first developed the material on body position presented here. They observed that more effective control of the patient's motion came when the therapist was in the line of the desired motion. As the therapist shifted position, the direction of the resistance changed and the patient's movement changed with it. From this knowledge they developed these guidelines for body position (G. Johnson and V. Saliba, unpublished hand-out, 1985):

– The therapist's body should be in line with the desired motion or force. To line up properly, the therapist's shoulders and pelvis face the direction of the motion. The arms and hands also line up with the motion. If the therapist cannot keep the proper body position, the hands and arms maintain alignment with the motion (Fig. 2.6).
– The resistance comes from the therapist's body while the hands and arms stay comparatively relaxed. By using body weight the therapist can give prolonged resistance without fatiguing. The relaxed hands allow the therapist to feel the patient's responses.

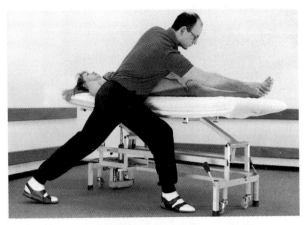

Fig. 2.6. Positioning of the therapist's body for the leg pattern flexion – abduction – internal rotation

2.5 Verbal (Commands)

The verbal command tells the patient what to do and when to do it. The therapist must always bear in mind that the command is being given to the *patient*, not to the body part being treated. Preparatory instructions need to be clear and concise, without unnecessary words. They may be combined with passive motion to teach the desired movement.

The timing of the command is important when using the *stretch reflex*. The initial command should come immediately before the stretch to coordinate the patient's conscious effort with the reflex response (Evarts and Tannji 1974). The action command should be repeated to urge greater effort or redirect the motion.

The volume with which the command is given can affect the strength of the resulting muscle contractions (Johansson et al. 1983): the therapist should give a louder command when a strong muscle contraction is desired and use a softer and calmer tone when the goal is relaxation or relief of pain.

The command is divided into three parts:

1. Preparation: readies the patient for action
2. Action: tells the patient to start the action
3. Correction: tells the patient how to correct and modify the action.

For example, the command for the lower extremity pattern flexion – adduction – external rotation with knee flexion might be "[preparation] ready, and [action] now pull your leg up and in; [correction] keep pulling your toes up" (to correct lack of dorsiflexion).

Fig. 2.7. Visual control

2.6 Vision

Using vision helps the patient control and correct his or her position and motion. Moving the eyes will influence both the head and body motion. For example, when a patient looks in the direction he or she wants to move, the head follows the eye motion; the head motion in turn will facilitate larger and stronger trunk motion (Fig. 2.7). The feedback from the visual sensory system can promote a more powerful muscle contraction. For example, when a patient looks at his or her arm or leg while exercising it a stronger contraction is achieved.

Eye contact between patient and therapist also provides another avenue of communication and helps to ensure cooperative interaction.

2.7 Traction and Approximation

Traction is the elongation of the trunk or an extremity. Knott, Voss, and their colleagnes theorized that the therapeutic effects of traction are due to stimulation of receptors in the joints (Knott and Voss 1968; Voss et al. 1985). Traction also acts as a stretch stimulus by elongating the muscles.

Traction is used to:

1. Facilitate motion, especially pulling and antigravity motions
2. Aid in elongation of muscle tissue when using the stretch reflex
3. Resist some part of the motion

Traction of the affected part is helpful when treating patients with joint pain.

The traction force should be applied gradually until the desired result is achieved. The traction should be maintained throughout the movement and combined with appropriate resistance.

Approximation is the compression of the trunk or an extremity. The muscle contractions following approximation are also thought to be due to stimulation of joint receptors (Knott and Voss 1968; Voss et al. 1985). Another possible reasons for the increased muscular response is muscle contractions to counteract the disturbance of position or posture caused by the approximation.

Approximation is used to:

1. Promote stabilization
2. Facilitate weight-bearing and the contraction of antigravity muscles
3. Resist some component of motion

Given gradually and gently, approximation may aid in treating painful and unstable joints.

There are two ways to apply the approximation:

1. Quick approximation: the force is applied quickly to elicit a reflex-type response
2. Slow approximation: the force is applied gradually up to the patient's tolerance

Regardless of whether the approximation is done quickly or slowly, the therapist must maintain the force and give resistance to the resulting muscular response. An appropriate command should be coordinated with the application of the approximation, for example "hold it" or "stand tall." The patient's joints should be properly aligned and in a weight-bearing position before the approximation is carried out.

2.8 Stretch

The *stretch stimulus* occurs when a muscle is elongated. The stimulus facilitates the elongated muscle, synergistic muscles at the same joint, and other associated synergistic muscles (Loofbourrow and Gellhorn 1948). Greater facilitation comes from lengthening all the synergistic muscles of a limb or the trunk. For example, elongation of the anterior tibial muscle facilitates that muscle and also facilitates the hip flexor – adductor – external rotator muscle group. If just the hip flexor – adductor – external rotator muscle group is elongated, the hip muscles and the anterior tibial muscle share the increased facilitation. If all the muscles of the hip and ankle are lengthened simultaneously, the excitability in those limb muscles increases further and spreads to the synergistic trunk flexor muscles.

The *stretch reflex* is elicited from muscles that are under tension, either from elongation or from contraction. The reflex has two parts: The first is a short-latency spinal reflex that produces little force and may not be of functional significance. The second part, called the *functional stretch response,* has a longer latency but produces a more powerful and functional contraction (Conrad and Meyer-Lohmann 1980; Chan 1984). To be effective as a treatment, the muscular contraction following the stretch must be resisted.

The strength of the muscular contraction following stretch is affected by the intent of the subject and, therefore, by prior instruction. Monkeys show changes in their motor cortex and stronger responses when they are instructed to resist the stretch. The same increase in response has been shown to happen in humans when they are told to resist a muscle stretch (Evarts and Tannji 1974; Chan 1984; Hammond 1956).

To elicit the reflex the therapist gives a rapid, gentle, elongating "tap" to muscles that are under tension. A preparatory command is given just before the stretch, for example: "now [preparation] – *pull* [action]" or "pull [preparation] – *harder* [action]." The timing and strength of the therapist's commands will influence the effectiveness of the patient's response to the stretch. For effective treatment, the therapist must resist the muscle contraction that results from the stretch. Because this reflex has a long latency, the therapist must wait for the muscular contraction to develop before giving resistance. Kabat (1947) stated that the stretch reflex may be the only way to get a very weak muscle to contract. How, why, and when to use the stretch reflex is described in Sect. 3.4.

2.9 Timing

Timing is the sequencing of motions. Normal movement requires a smooth sequence of activity, and coordinated movement requires precise timing of that sequence. Functional movement requires continued, coordinated motion until the task is accomplished.

Normal timing of most coordinated and efficient motions is from distal to proximal. The evolution of control and coordination during development proceeds from cranial to caudal and from proximal to distal (Jacobs 1967). In infancy the arm determines where the hand goes, but after the grasp matures the hand directs the course of the arm movements (Halvorson 1931). The small motions that adults use to maintain standing balance proceed from distal (ankle) to proximal (hip and trunk) (Nashner 1977). To restore normal timing of motion may become a goal of the treatment.

Timing for emphasis involves changing the normal sequencing of motions to emphasize a particular muscle or a desired activity. Kabat (1947) wrote that prevention of motion in a stronger synergist will redirect the energy of that contraction into a weaker muscle. This alteration of timing stimulates the proprioceptive reflexes in the muscles by resistance and stretch. The best results come when the strong muscles score at least "good" in strength (Manual Muscle Test grade 4; Partridge 1954).

There are two ways the therapist can alter the normal timing for therapeutic purposes (Figs. 2.8, 2.9):

1. By preventing all the motions of a pattern except the one that is to be emphasized.
2. By resisting an *isometric* or *maintained* contraction of the strong motions in a pattern while exercising the weaker muscles. This resistance to the static contraction *locks in* that segment, so resisting the contraction is called "locking it in."

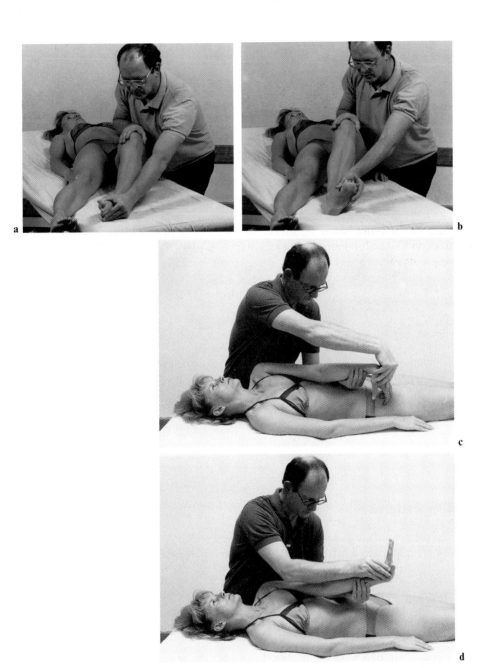

Fig. 2.8 a–d. Timing for emphasis by preventing motion. **a, b** Leg pattern flexion – abduction – internal rotation. The strong motions of the hip and knee are blocked and dorsiflexion – eversion of the ankle exercised using repeated stretch. **c, d** Arm pattern flexion – abduction – external rotation. The stronger shoulder motions are blocked while exercising radial extension of the wrist

13

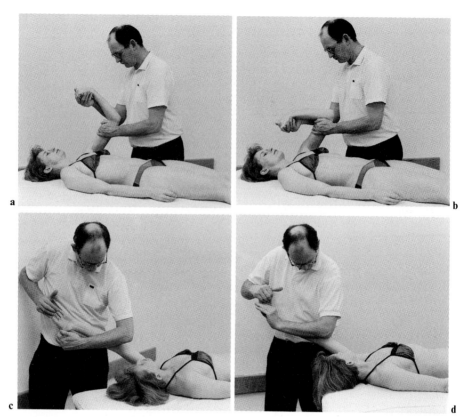

Fig. 2.9 a–d. Timing for emphasis using isometric contractions of strong muscles. *a, b* Exercising elbow flexion using the pattern flexion – adduction – external rotation. *c, d* Exercising finger flexion using the pattern extension – adduction – internal rotation

2.10 Patterns

The patterns of facilitation may be considered one of the basic procedures of PNF. For greater clarity we discuss and illustrate them together in Chap. 5.

References

Chan CWY (1984) Neurophysiological basis underlying the use of resistance to facilitate movement. Physiother Canada 36 (6): 335–341

Conrad B, Meyer-Lohmann J (1980) The long-loop transcortical load compensating reflex. TINS 3: 269–272

Evarts EV, Tannji J (1974) Gating of motor cortex reflexes by prior instruction. Brain Res 71: 479–494

Fischer E (1967) Factors affecting motor learning. Am J Phys Med 46 (1): 511–519

Gellhorn E (1947) Patterns of muscular activity in man. Arch Phys Med 28: 568–574

Gellhorn E (1949) Proprioception and the motor cortex. Brain 72: 35–62

Halvorson HM (1931) An experimental study of prehension in infants by means of systematic cinema records. Genet Psychol Monogr 10: 279–289. Reprinted in: Jacobs MJ (1967) Development of normal motor behavior. Am J Phys Med 46 (1): 41–51

Hammond PH (1956) The influences of prior instruction to the subject on an apparently involuntary neuromuscular response. J Physiol (Lond) 132: 17P–18P

Hislop HH (1960) Pain and exercise. Phys Ther Rev 40 (2): 98–106

Jacobs MJ (1967) Development of normal motor behavior. Am J Phys Med 46 (1): 41–51

Johansson CA, Kent BE, Shepard KF (1983) Relationship between verbal command volume and magnitude of muscle contraction. Phys Ther 63 (8): 1260–1265

Kabat H (1947) Studies on neuromuscular dysfunction, XI: New principles of neuromuscular reeducation. Perm Found Med Bull 5 (3): 111–123

Kabat H (1961) Proprioceptive facilitation in therapeutic exercise. In: Licht S Johnson EW (eds) Therapeutic exercise, 2nd ed. Waverly, Baltimore

Knott M, Voss DE (1968) Proprioceptive neuromuscular facilitation: patterns and techniques, 2nd ed. Harper and Row, New York

Loofbourrow GN, Gellhorn E (1949) Proprioceptive modification of reflex patterns. J Neurophysiol 12: 435–446

Loofbourrow GN, Gellhorn E (1948) Proprioceptively induced reflex patterns. Am J Physiol 154: 433–438

Nashner LM (1977) Fixed patterns of rapid postural responses among leg muscles during stance. Exp Brain Res 30: 13–24

Partridge MJ (1954) Electromyographic demonstration of facilitation. Phys Ther Rev 34 (5): 227–233

Sherrington C (1947) The integrative action of the nervous system, 2nd edn. Yale University Press, New Haven

Voss DE, Ionta M, Meyers B (1985) Proprioceptive neuromuscular facilitation: patterns and techniques, 3rd ed. Harper and Row, New York

Webster's ninth new collegiate dictionary (1984) Merriam-Webster, Springfield

Further Reading

General

Griffin JW (1974) Use of proprioceptive stimuli in therapeutic exercise. Phys Ther 54 (10): 1072–1079

Payton OD, Hirt S, Newton RA (eds) (1977) Scientific basis for neuro-physiologic approaches to therapeutic exercise: an anthology. Davis, Philadelphia

Resistance, Irradiation and Reinforcement

Hellebrandt FA (1958) Application of the overload principle to muscle training in man. Arch Phys Med Rehabil 37: 278–283

Hellebrandt FA, Houtz SJ (1956) Mechanisms of muscle training in man: experimental demonstration of the overload principle. Phys Ther 36 (6): 371–383

Hellebrandt FA, Houtz SJ (1958) Methods of muscle training: the influence of pacing. Phys Ther 38: 319–322

Hellebrandt FA, Waterland JC (1962) Expansion of motor patterning under exercise stress. Am J Phys Med 41: 56–66

Moore JC (1975) Excitation overflow: an electromyographic investigation. Arch Phys Med Rehabil 56: 115–120

Stretch

Burg D, Szumski AJ, Struppler A, Velho F (1974) Assessment of fusimotor contribution to reflex reinforcement in humans. J Neurol Neurosurg Psychiatry 37: 1012–1021

Cavagna GA, Dusman B, Margaria R (1968) Positive work done by a previously stretched muscle. J Appl Physiology 24 (1): 21–32

Chan CWY, Kearney RE (1982) Is the functional stretch response servo controlled or preprogrammed? Electroenceph Clin Neurophysiol 53: 310–324

Ghez C, Shinoda Y (1978) Spinal mechanisms of the functional stretch reflex. Exp Brain Res 32: 55–68

3 Techniques

The goal of the PNF techniques is to promote functional movement through facilitation, inhibition, strengthening, and relaxation of muscle groups. The techniques use concentric, eccentric, and static muscle contractions combined with properly graded resistance and suitable facilitory procedures and are combined and adjusted to fit the needs of each patient.

Example: Increasing the range of motion and strengthening the muscles in the newly gained range of motion.
A relaxation technique such as *contract-relax* is used to increase range of motion. This is followed by a facilitory technique such as *slow reversals* or *combination of isotonics* to increase the strength and control in the newly gained range of motion.
Example: Relieving muscle fatigue during strengthening exercises.
After using a strengthening technique such as *repeated stretch,* the technique of *slow reversals* is used to relieve fatigue. The stretch reflex permits muscles to work for longer without fatiguing. Alternating contractions of the antagonistic muscles relieve the fatigue that follows repeated exercise of one group of muscles.

We have grouped the PNF techniques below so that those with similar functions or actions are together. Where new terminology is used the name describes the activity or type of muscle contraction involved and when the terminology differs from that used by Knott and Voss (1968) both names are given. *Reversal of antagonists* is a general class of techniques in which the patient first contracts the agonistic muscles then contracts their antagonists without pause or relaxation. Within that class, *dynamic reversal of antagonist* is an isotonic technique where the patient first moves in one direction and then in the opposite without stopping. *Rhythmic stabilization* involves isometric contractions of the antagonistic muscle groups. In this technique, *no* motion is intended by either the patient or the therapist. We use both reversal techniques to increase strength and range of motion. *Rhythmic stabilization* works to increase the patient's ability to stabilize or hold a position as well.[1]

In presenting each technique we give a short characterization, the goals, uses, and any contraindications. The full description, examples, and ways in which the technique may be modified follow.

---

[1] G. Johnson and V. Saliba were the first to use the terms "stabilizing reversal of antagonists," "dynamic reversal of antagonist," "combination of isotonics," and "repeated stretch" in an unpublished course hand-out at the Institute of Physical Art (1979).

The techniques described are:

1. Rhythmic initiation
2. Combination of isotonics (G. Johnson and V. Saliba, unpublished hand-out, 1988) (reversal of agonists; Sullivan et al. 1982)
3. Reversal of antagonists
   a) Dynamic reversal of antagonists (incorporates slow reversal)
   b) Stabilizing reversal
   c) Rhythmic stabilization
4. Repeated stretch (repeated contraction)
   a) Repeated stretch from beginning of range
   b) Repeated stretch through range
5. Contract-relax
6. Hold-relax

3.1 Rhythmic Initiation

Characterization
Rhythmic motion done through the desired range, starting with passive motion and progressing to active resisted movement.

Goals
- Aid in initiation of motion
- Improve coordination and sense of motion
- Normalize the rate of motion, either increasing or decreasing it
- Teach the motion
- Help the patient to relax

Indications
- Difficulties in initiating motion
- Movement too slow or too fast
- Uncoordinated or dysrhythmic motion
- General tension

Description
- The therapist starts by moving the patient passively through the range of motion, using the speed of the verbal command to set the rhythm.
- The patient is asked to begin working actively in the desired direction. The return motion is done by the therapist.
- The therapist resists the active movement, maintaining the rhythm with the verbal commands.

Example
Trunk flexion in a sitting position:

- Move the patient passively into trunk flexion and then back to the upright position: "Let me move you forward. Good, now let me move you back and then forward again."

18

- When the patient is relaxed and moving easily, ask for active assisted motion: "Help me a little coming forward. Now relax and let me bring you back."
- Then begin resisting the motion: "Push forward towards me. Let me bring you back. Now push forward towards me again."

Modifications
- The technique can be finished by using eccentric as well as concentric muscle contractions (combination of isotionics).
- The technique may be done with active motion in both directions (reversal of antagonists).

3.2 Combination of Isotonics

Characterization
Concentric, eccentric, and stabilizing contractions of one group of muscles (agonists) without relaxation.

Goals
- Active control of motion
- Coordination
- Increase the active range of motion
- Strengthen
- Functional training in eccentric control of movement

Indications
- Decreased eccentric control
- Lack of coordination or ability to move in a desired direction
- Decreased active range of motion
- Lack of active motion in the middle of the range

Description
- The therapist resists the patient's moving actively through a range of motion (concentric contraction).
- At the end of the motion the therapist tells the patient to stay in that position (stabilizing contraction).
- When stability is attained the therapist tells the patient to allow the part to be moved slowly back to the starting position (eccentric contraction).
- Note: The eccentric muscle contraction may come before the concentric contraction.

Example
Trunk flexion in a sitting position:
- Resist the patient's concentric contraction into trunk flexion: "Push forward towards me."
- At the end of the patient's active range of motion, tell the patient to stabilize in that position: "Stop, stay there, don't let me push you back."

- After the patient is stable, move the patient back to the original position while he or she maintains control with an eccentric contraction of the trunk flexor muscles: "Now let me push you back, but slowly."

Modifications
- The technique may be combined with reversal of antagonists.
- The technique can start at the end of the range of motion and begin with eccentric contractions.
- One type of muscle contraction can be changed to another before completing the full range of motion.
- A change can be made from the concentric to the eccentric muscle contraction without stopping or stabilizing.

3.3 Reversal of Antagonists

3.3.1 Dynamic Reversals (Incorporates Slow Reversal)

Characterization
Active motion changing from one direction to the opposite without pause or relaxation.

Goals
- Increase active range of motion
- Increase strength
- Develop coordination (smooth reversal of motion)
- Prevent or reduce fatigue

Indications
- Weakness of the agonistic muscles
- Decreased ability to change direction of motion
- Exercised muscles begin to fatigue

Description
- The therapist resists the patient's moving in one direction.
- As the end of the desired range of motion approaches, the therapist reverses the grip on the distal portion of the moving segment.
- As the patient reaches the end of his or her active range of motion, the therapist gives the command to reverse direction, without relaxation, and resists the new motion at the distal part.
- When the patient begins moving in the opposite direction the therapist reverses the proximal grip so all resistance opposes the new direction.
- The reversals may be done as often as necessary.

Examples
Reversing upper extremity motion from flexion to extension:

- Resist the desired pattern of arm flexion: "Wrist back and lift your arm up."

- As the patient's arm approaches the end of the range, move the hand that was resisting proximally (on the arm or scapula) so that it can resist the distals component (patient's hand) during the reverse motion: "Now squeeze my hand and pull your arm down."
- As the patient starts to move in the new direction, move your other hand so that it now resists the proximal part of the pattern.

Reversing lower extremity motion from flexion to extension:

- Resist the desired pattern of lower extremity flexion: "Foot up and lift your leg up."
- As the patient's leg approaches the end of the range, slide the hand that was resisting on the dorsum of the foot to the plantar surface so that it can resist the patient's foot during the reverse motion: "Now push your foot down and kick your leg down."
- As the patient starts to move in the new direction, move the proximal hand so that it now resists the new direction of motion.

Modifications
- Instead of moving through the full range, the change of direction can be used to emphasize a particular range of the motion.
- The speed used in one or both directions can be varied.
- The technique can begin with small motions in each direction, increasing the range of motion as the patient's skill increases.
- The range of motion can be decreased in each direction until the patient is stabilized in both directions.
- The patient can be instructed to hold his or her position or stabilize at any point in the range of motion or at the end of the range. This can be done before and after reversing direction.
- The technique can begin with the stronger direction to gain irradiation into the weaker muscles after reversing.
- A reversal can be done whenever the agonistic muscles begin to fatigue.

3.3.2 Stabilizing Reversals

Characterization
Alternating isotonic contractions opposed by enough resistance to prevent motion.

Goals
- Increase stability and balance
- Increase muscle strength

Indications
- Decreased stability
- Weakness
- Patient unable to contract muscle isometrically

Description

- The therapist gives resistance to the patient in one direction while asking the patient to oppose the force, no motion is allowed.
- When the patient is fully resisting the force the therapist moves one hand and begins to give resistance in another direction.
- After the patient responds to the new resistance the therapist moves the other hand to resist the new direction.

Example

Trunk stability

- Resist the patient's trunk flexion: "Don't let me push you backward."
- When the patient is contracting his or her trunk flexor muscles, maintain the resistance with one hand while moving your other hand to resist the patient's trunk extension: "Now don't let me pull you forward."
- As the patient responds to the new resistance, move the hand that was still resisting trunk flexion to resist trunk extension.
- Reverse directions as often as needed to be sure the patient is stable: "Now don't let me push you. Now switch again, and don't let me pull you."

Modifications

- The technique can begin with slow reversals and progress to smaller ranges until the patient is stabilizing.
- The stabilization can be begun with the stronger muscle groups to irradiate into the weaker muscles.
- The resistance may be moved around the patient so that all muscle groups work.
- The speed of the reversal may be increased or decreased.

3.3.3 Rhythmic Stabilization

Characterization

Alternating isometric contractions against resistance, no motion intended.[2]

Goals

- Increase active and passive range of motion
- Increase strength
- Increase stability and balance
- Decrease pain

Indications and contraindications

Indications
- Limited range of motion
- Pain, particularly when motion is attempted

---

[2] In the first and second editions of *Proprioceptive neuromuscular facilitation,* Knott and Voss describe this technique as resisting alternately the agonistic and antagonistic patterns without relaxation. In the third edition (1985), Voss et al. describe resisting the agonistic pattern distally and the antagonistic pattern proximally.

- Joint instability
- Weakness in the antagonistic muscle group
- Decreased balance

Contraindications
- Cerebellar involvement (Kabat 1950)
- Patient unable to follow instructions due to age, language difficulty, cerebral dysfunction

Description
- The therapist resists an isometric contraction of the agonistic muscle group. The patient maintains the position of the part without trying to move.
- The resistance is increased slowly as the patient builds a matching force.
- When the patient is responding fully, the therapist moves one hand to begin resisting the antagonistic motion at the distal part.
- The new resistance is built up slowly. As the patient responds the therapist moves the other hand to resist the antagonistic motion also.
- The reversals are repeated as often as needed.

Example
Trunk stability

- Resist an isometric contraction of the patient's trunk flexor muscles: "Stay still, match my resistance in front."
- Next, take all the anterior resistance with your left hand and move your right hand to resist trunk extension: "Now start matching me in back, hold it."
- As the patient responds to the new resistance, move your left hand to resist trunk extension: "Stay still, match me in back."
- The direction of contraction may be reversed as often as necessary to reach the chosen goal: "Now hold in front again. Stay still. Now start matching me in the back."

Modifications
- The technique can begin with the stronger group of muscles for facilitation of the weaker muscle group.
- The stabilizing activity can be followed by a strengthening technique for the weak muscles.
- To increase the range of motion the stabilization may be followed by asking the patient to move further into the restricted range.
- For relaxation the patient may be asked to relax all muscles at the end of the technique.
- To gain relaxation without pain the technique may be performed on muscles distant from the painful area.

3.4 Repeated Stretch (Repeated Contractions)

3.4.1 Repeated Stretch from Beginning of Range

Characterization

Stretch reflex elicited from muscles under the *tension of elongation.*
(*Note:* Only muscles should be under tension; care must be taken not to stretch the joint structures.)

Goals
- Facilitate initiation of motion
- Increase active range of motion
- Increase strength
- Prevent or reduce fatigue
- Guide motion in the desired direction

Indications and contraindications

Indications
- Weakness
- Inability to initiate motion due to weakness or rigidity
- Fatigue
- Decreased awareness of motion

Contraindications
- Joint instability
- Pain
- Unstable bones due to fracture or osteoporosis
- Damaged muscle or tendon

Description
- The therapist gives a preparatory command while elongating fully all the muscles in the pattern. Particular attention is paid to the rotation.
- A quick "Tap" is given to lengthen the muscles further and evoke the stretch reflex.
- At the same time as the stretch the therapist gives a command to link the patient's voluntary effort to contract the stretched muscles with the reflex response.
- The resulting muscle contraction, reflex and voluntary, is resisted.

Example

Stretch of the pattern of flexion–abduction–internal rotation
- Rotate the patient's lower extremity into external rotation and then place the foot in plantar flexion-inversion and the hip into full extension, adduction, and external rotation.
- When all the muscles of the flexion–abduction–internal rotation pattern are taut, give the preparatory command "Now!" while quickly elongating (stretching) all the muscles farther.
- Immediately after the stretch, give the command "Pull up and out."

– When you feel the patient's muscles contract, give resistance to the entire pattern.

Modifications
– The technique may be repeated from the beginning of the range as soon as the contraction weakens or stops.
– The resistence may be modified so that only some of the motions are allowed to occur (timing for emphasis). For example, the therapist may prevent any hip motion from occurring while resisting the ankle dorsiflexion and eversion through its range.

3.4.2 Repeated Stretch Through Range

Characterization
Stretch reflex elicited from muscles under the *tension of contraction* (Fig. 3.1).

Goals
– Increase active range of motion
– Increase strength
– Prevent or reduce fatigue
– Guide motion in the desired direction

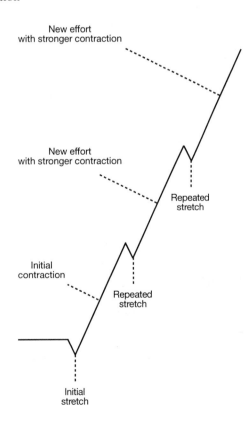

Fig. 3.1. Repeated stretch through the range

Indications and contraindications

Indications
- Weakness
- Fatigue
- Decreased awareness of desired motion

Containdications
- Joint instability
- Pain
- Unstable bones due to fracture or osteoporosis
- Damaged muscle or tendon
- Insufficient strength to maintain the contraction of the muscles

Description
- The therapist resists a pattern of motion so all the muscles are contracting and tense.
- The therapist gives a preparatory command to coordinate the stretch reflex with a new, increased effort by the patient.
- At the same time the therapist slightly elongates (stretches) the muscles by momentarily giving too much resistance.
- A new and stronger muscle contraction is asked for and resisted.
- The stretch may be repeated to strengthen or redirect the motion as the patient moves through the range.
- The patient must be allowed to move before the next stretch reflex is given.
- The patient must not relax or reverse direction during the stretch.

Example
Repeated stretch of the lower extremity pattern of flexion–abduction– internal rotation

- Resist the patient's moving his or her lower extremity into flexion–abduction–intenal rotation: "Foot up, pull your leg up and out."
- Give a preparatory command–"Now!"–while slightly over-resisting the motion so that you pull the patient's leg a short distance back in the direction of extension–adduction–external rotation. The patient must maintain the contraction of the stretched muscles.
- Give the command "Pull again, harder" immediately after the stretch.
- Now give appropriate resistance to the increased contraction that follows the re-stretch of the muscles.
- Repeat the stretch and resistance as you feel the patient's strength decreasing.
- Repeat the stretch if you feel the patient start to move in the wrong direction.

Modifications
- The therapist may ask for a stabilizing contraction of the pattern before re-stretching the muscles: "Hold your leg here, don't let me pull it down. Now, pull it up harder."
- The therapist may resist a stabilizing contraction of the stronger muscles in the

26

pattern while re-stretching and resisting the weaker muscles (timing for emphasis). For example, the hip motion can be "locked in" by resisting a stabilizing contraction of those muscles: "Hold your hip there." The ankle motion of dorsiflexion and eversion is re-stretched and the new contraction resisted through range: "Pull your ankle up and out harder."

3.5 Contract-Relax

Characterization
Resisted isotonic contraction of the restricting muscles followed by relaxation and movement into the increased range.

Goal
– Increased passive range of motion

Indication
– Decreased passive range of motion

Description
– The therapist or the patient moves the joint or body segment to the end of the passive range of motion. Active motion is preferred and the therapist may resist it.
– The therapist asks the patient for a strong contraction of the restricting muscle or pattern (antagonists).
– Enough motion is allowed to happen for the therapist to be certain that all the desired muscles, particularly the rotators, are contracting.
– After sufficient time (at least 5 seconds), the therapist tells the patient to relax.
– Both the patient and the therapist relax.
– The joint or body part is repositioned, either actively by the patient or passively by the therapist, to the new limit of the passive range. Active motion is preferred and may be resisted.
– The technique is repeated until no more range is gained.
– Active resisted exercise of the agonistic and antagonistic muscles is done in the new range of motion.

Example
Increasing the range of shoulder flexion, abduction, and external rotation.

– The patient moves the arm to the end of his or her range of flexion–abduction–external rotation: "Open your hand and lift your arm up as high as you can."
– Resist an isotonic contraction of the pattern of extension–adduction–internal rotation: "Squeeze my hand and pull your arm down and across. Keep turning your hand down."
– Allow enough motion to occur for both you and the patient to know that all the muscles in the pattern, particularly the rotators, are contracting: "Keep pulling your arm down."
– After resisting the contraction for at least 5 seconds, both you and the patient relax: "Relax, let everything go loose."

- Now, resist the patient's motion into the newly gained range: "Open your hand and lift your arm up farther."
- When no more range is gained, exercise the agonistic and antagonistic patterns, either in the new range or throughout the entire range of motion: "Squeeze and pull your arm down; now open your hand and lift your arm up again."

Modifications
- The technipue can be done using contraction of the agonistic muscles: "Don't let me push your arm down, keep pushing up."
- The patient is asked to move immediately into the desired range without any relaxation.
- Alternating contractions (reversals) of agonistic and antagonistic muscles may be done: "Keep your arm still, don't let me pull it up. Now don't let me push your arm down."

3.6 Hold-Relax

Characterization
Resisted isometric contraction followed by relaxation.

Goals
- Increase passive range of motion
- Decrease pain

Indications and Contraindication

Indications
- Decreased passive range of motion
- The patient's isotonic contractions are too strong for the therapist to control
- Pain

Contraindication
- The patient is unable to do an isometric contraction

Description

For increasing range of motion
- The therapist or patient moves the joint or body segment to the end of the passive or pain-free range of motion. Active motion is preferred. The therapist may resist if it does not cause pain.
- The therapist asks for an isometric contraction of the restricting muscle or pattern (antagonists) with emphasis on rotation.
- The resistance is increased slowly.
- No motion is intended by either the patient or the therapist.
- After holding the contraction for enough time (at least 5 seconds) the therapist asks the patient to relax.
- Both the therapist and the patient relax gradually.

- The joint or body part is repositioned either actively or passively to the new limit of range. Active motion is preferred if it is pain-free. The motion may be resisted if that does not cause pain.
- Repeat all steps in the new limit of range.

For decreasing pain

Direct treatment
- The patient is in a position of comfort.
- The therapist resists an isometric contraction of muscles affecting painful segment.

Indirect treatment
- The patient is in a position of comfort.
- The therapist resists isometric contractions of muscles distant from the painful segment.
- The resistance is built up slowly and remains at a level below that which causes pain.
- During relaxation the resistance decreases slowly.

Example
Indirect treatment for decreasing pain in the right shoulder.

- The patient lies with his or her right arm supported in a comfortable position and the right elbow flexed.
- Hold the patient's right hand and ask for an isometric contraction of the ulnar flexor muscles of the wrist: "Keep your hand and wrist right there. Match my resistance."
- Resist an isometric contraction of the ulnar wrist flexors and forearm pronator muscles. Build the resistance up slowly and keep it at a pain-free level for at least 5 seconds: "Keep holding, match my resistance."
- While maintainig the resistance, monitor muscle activity in the patient's right shoulder, particularly the internal rotation.
- Both you and the patient relax slowly and completely: "Now let go slowly all over."
  Repeat the technique in the same position to gain more relaxation, or move the forearm into more supination or pronation to change the effect on the shoulder muscles.

Modifications
- The technique may be done with contraction of the agonistic muscles, in this case resisting an isometric contraction of the radial extensor muscles of the wrist and the forearm supinator muscles.
- Alternating isometric contractions or *rhythmic stabilization* may be done.
- If the patient is unable to do an isometric conatraction, carefully controlled stabilizing contractions may be used. The therapist's resistance and the patient's effort must stay at a level that does not cause pain.

3.7 Summary of PNF Techniques and Their Goals

The PNF techniques which can be used to achieve a particular goal are outlined below.

1. Initiate motion
– Rhythmic initiation
– Repeated stretch from beginning of range

2. Learn a motion
– Rhythmic initiation
– Combination of isotonics
– Repeated stretch from beginning of range
– Repeated stretch through range

3. Change rate of motion
– Rhythmic initiation
– Dynamic reversals
– Repeated stretch from beginning of range
– Repeated stretch through range

4. Increase strength
– Combination of isotonics
– Dynamic reversals
– Rhythmic stabilization
– Stabilizing reversals
– Repeated stretch from beginning of range
– Repeated stretch through range

5. Increase stability
– Combination of isotonics
– Stabilizing reversals
– Rhythmic stabilization

6. Increase coordination and control
– Rhythmic initiation
– Combination of isotonics
– Dynamic reversals
– Stabilizing reversals
– Rhythmic stabilization
– Repeated stretch from beginning of range

7. Increase endurance
– Dynamic reversals
– Stabilizing reversals
– Rhythmic Stabilization

- Repeated stretch from beginning of range
- Repeated stretch through range

8. Increase range of motion
- Dynamic reversals
- Stabilizing Reversals
- Rhythmic stabilization
- Repeated stretch form beginning of range
- Contract-relax
- Hold-relax

9. Relaxation
- Rhythmic initiation
- Rhythmic stabilization
- Hold-relax

10. Decrease pain
- Rhythmic stabilization (or stabilizing reversals)
- Hold-relax

References

Kabat H (1950) Studies on neuromuscular dysfunction, XII: Rhythmic stabilization; a new and more effective technique for treatment of paralysis through a cerebellar mechanism. Perm Found Med Bull 8 (1): 9–19

Knott M, Voss DE (1956) Proprioceptive neuromuscular facilitation: Patterns and techniques. Harper and Row, New York

Knott M, Voss DE (1968) Proprioceptive neuromuscular facilitation: Patterns and techniques, 2nd ed. Harper and Row, New York

Sullivan PE, Markos PD, Minor MAD (1982) An integrated approach to therapeutic exercise. Reston, Virginia

Voss DE, Ionta M, Myers BT (1985) Proprioceptive neuromuscular facilitation, 3rd ed.

Further Reading

Markos PD (1979) Ipsilateral and contralateral effects of proprioceptive neuromuscular facilitation techniques on hip motion and electro-myographic activity. Phys Ther 59 (11): 1366–1373

Moore M, Kukulka C (1988) Depression of H reflexes following voluntary contraction. Phys Ther 68: 862

Rose-Jacobs R, Gilberti N (1984) Effect of PNF and Rood relaxation techniques on muscle length. Phys Ther 64: 725

Sady SP, Wortman M, Blanke D (1982) Flexibility training: ballistic, static or proprioceptive neuromuscular facilitation? Arch Phys Med Rehabil 63: 261–263

Tanigawa MC (1972) Comparison of the hold-relax procedure and passive mobilization on increasing muscle lenght. Phys Ther 52: 725–735

4 Patient Treatment

Our treatment seeks to help each patient gain the highest level of function possible. An effective treatment depends on our doing a complete and accurate evaluation[1] to identify the patient's areas of function and dysfunction. On the basis of this evaluation we set general and specific goals and then design a treatment plan to achieve them. Continuous assessment[2] guides us in adjusting the treatment as the patient progresses.

4.1 Evaluation

Working within the PNF philosophy, we look first for the patient's areas of *function*. We will use these strong areas to construct an effective treatment for each problem. Next we note the patient's *general (functional) problems*. Last, the specific *dysfunctions* which are causing the general problems are identified.

1. Function
   a) Pain-free
   b) Strong
   c) Able to move and stabilize
   d) Motion is controlled and coordinated

Example: Mr. Brown has a right hemiplegia:

1. His left side is strong and pain-free.
   We can use irradiation and reinforcement from the left side to treat the right side.
2. He can roll from supine to prone and back using his left arm and leg.
   He can change position in bed (function). We can use rolling to strengthen his trunk and right side.
3. He can go from supine to sitting and back in less than 1 minute.
   He can get to sitting in a reasonable amount of time (function). We can resist this activity to increase trunk and right side strength.
4. His sitting balance (static and dynamic) is good.
   He is functional when sitting. He will be able to exercise in a sitting position.
5. He can get up from a chair without help.

---

[1] To evaluate: to identify the patient's areas of function and dysfunction.
[2] To assess: to measure or judge the result of a treatment procedure.

He is functional in changing position. We can resist this activity to increase strength in his right leg.
6. His standing balance is good.
   He will be able to exercise in the standing position.
7. He can walk with only stand-by assistance with a cane and an ankle–foot orthosis on the right leg.
   We can use resistance in walking to increase the strength of his trunk and legs.
8. He takes steps with the right leg and bears weight on it.
   He has some strength and control in the right leg. He can take resistance when exercising.
9. He has normal sensation and position awareness in his right hip and knee.
   He does not have to use vision and control the motion of his right hip or knee during exercise or for function.

2. Dysfunction

a) General (functional) loss
   I)   Static: loss of the ability to maintain a position
   II)   Dynamic: loss of the ability to move or control motion

Example: Mr. Brown's general problems:
1. He has no voluntary motion of his right arm.
2. He has shoulder pain when his right arm is moved above 120° flexion.
3. When walking he does not extend his right hip past neutral in stance.
4. The spasticity in the right ankle increases when he walkes.
5. He has difficulty pronouncing words distinctly.
6. After eating there is food trapped between his teeth and right cheek.
7. When sitting and standing, there is a shortening of the right side of the trunk and weight is born on the left side.

b) Specific deficits (the reasons for the functional losses)
   I)   Pain
   II)   Decreased range of motion: muscle tightness or shortening *or* joint restrictions
   III) Weakness
   IV)   Loss of sensation, proprioception
   V)   Deficit in sight, hearing
   VI)   Deficient motor control
   VII) Lack of endurance

Example: Mr. Brown's specific problems:

1. The decreased range of motion of the shoulder is due to pain; the end-feel is empty (possibly due to decreased range of motion in scapula, see 8a): general problem 2.
2. He shows no awareness of his right arm unless he feels pain in it (decreased sensation or proprioception): general problem 1.
3. He has generalized weakness throughout his right leg: general problem 3.

4. When supine, his right hip does not extend actively or passively beyond neutral (restriction is due to hip flexor muscle tightness and decreased pelvic mobility: General problem 3.
5. He has difficulty combining hip extension with knee flexion and hip flexion with knee extension in the right leg in all positions (evidence of diminished selective control of these muscles): general problem 3.
6. The passive range of right ankle dorsiflexion is $-5°$ (restriction is due to muscle tightness). He has no voluntary ankle motion: general problem 4
7. He has generalized weakness in the muscles on the lower right side of his face: general problems 5 and 6.
8a. The scapula does not have the full passive range of motion in any pattern. The pelvis is limited in both anterior and posterior depression patterns: general problems 2, 3, and 7.
8b. He has poor active motion in both the scapula and pelvis: general problems 1, 2, 3, and 7.

4.2 Treatment Goals

After doing the evaluation, we set general and specific treatment goals. *General goals* are expressed as functional activities. For example,

1. A patient who has general mobility problems can move from supine to sitting in bed without help.
2. A patient who has a problem with his right knee can run one mile (1,6 km) in less than 6 minutes without pain in the knee.
3. A patient who has had a stroke can walk 25 feet ($\approx$ 8 m) in 2 minutes using a cane and an ankle–foot orthosis.

These goals are not limits. We must change them as the patient improves.

Specific goals are set for each treatment activity and treatment session. Examples for the general problems are:

1. The patient can roll from supine to sidelying and back 10 times in 1 minute. (Begin treatment with general mobility and the preliminary motions needed to get into a sitting position.)
2. The patient can hold a one-leg bridging position on the right leg with the left leg extended for 30 seconds. (Begin treatment with limited weight-bearing on the right leg.)
3. The patient can shift weight from the right to the left ischial tuberosity while sitting without any support. (Begin treatment with weight shifting in a stable position.)

4.3 Treatment Design

The therapist must design a treatment to meet each patient's specific needs and functional goals. PNF uses muscle contractions to affect the body. If muscle contractions are not appropriate for the patient's condition or if their use does not achieve the desired goals, the therapist should use other methods. Modalities such as heat and cold, passive joint motion, and soft tissue mobilization may be combined with PNF for effective treatment.

Specific Patient Needs. The therapist lists the patient's needs, such as to:
1. Decrease pain
2. Increase range of motion
3. Increase strength, coordination, and control of motion
4. Develop a proper balance between motion an stability
5. Increase endurance

Designing the Treatment. The therapist designs a treatment to meet the patient's needs. Factors to be considered include:
1. Choosing to use direct or indirect treatment
2. Choosing appropriate activities
   a) Movement/stability
   b) Types of muscle contractions
3. Choosing the techniques and procedures
4. Selecting the patterns and combinations of patterns
5. Deciding on the best position for the patient, considering
   a) Effect of gravity
   b) Effect on two-joint muscles
   c) Reflex facilitation
   d) Use of vision

4.4 Direct and Indirect Treatment

Many studies have shown the effectiveness of indirect treatment that begins on strong and pain-free parts of the body. Hellebrandt et al. (1947) reported the development of muscle tension in unexercised parts of the body during and after maximal exercise of one limb. Other experiments have described electromyographic (EMG) activity in the agonistic and antagonistic muscles of the contralateral upper or lower extremity during resisted isotonic and isometric exercise (Devine et al. 1981; Pink 1981; Moore 1975). The trunk musculature can also be exercised indirectly. For example, the abdominal muscles contract synergistically when a person raises his arm. This activity occurs in normal subjects and in patients suffering from central nervous system disorders as well (Angel and Eppler 1967). An increased passive range of motion can be gained indirectly by using *contract-relax* on uninvolved areas of the body (Markos 1979)

To give the patient maximum benefit from indirect treatment the therapist resists strong movements or patterns. Maximum strengthening occurs when the patient's strong limbs work in combination with the weak ones.

When pain is a presenting symptom, treatment focuses on *pain-free* areas of the body. Using carfully guided and controlled irradiation the therapist can treat the affected limb or joint without risk of increasing the pain or injury.

4.4.1 Direct Treatment

Direct treatment may involve:
1. Use of treatment techniques on the affected limb, muscle, or motion.
   Example: To gain increased range in the shoulder motions of flexion, abduction, and external rotation, the therapist treats the involved shoulder using the technique *contract-relax* on the tight pectoralis major muscle.
2. Directing the patient's attention to stabilizing or moving the affected segment.
   Example: While the patient stands on the involved leg, the therapist gives approximation through the pelvis to facilitate weight-bearing.

4.4.2 Indirect Treatment

Indirect treatment may involve:
1. Use of the techniques on an unaffected or less affected part of the body. The therapist directs the irradiation into the affected area to achieve the desired results.
   Example: To gain range in shoulder flexion, abduction and external rotation, the therapist resists an isometric contraction of the ulnar wrist flexor and the pronator muscles of the affected arm. After resisting the contraction, the therapist and the patient relax. This use of *hold-relax* will produce a contraction and relaxation of the ipsilateral pectoralis major muscle. The treated arm need not be moved but may remain in a position of comfort.
2. Directing the patient's attention and effort toward working with the less affected parts of the body.
   Example: While the patient sits with both feet on the floor, the therapist resists the "lifting" pattern (trunk extension) on the side of the involved lower extremity. This produces contraction of the extensor muscles in the lower extremities and increased weight-bearing through the ipsilateral ischial tuberosity and the foot.

4.5 Assessment

The process of patient evaluation and treatment assessment is continuous. By assessing the results after each treatmant, the therapist can determine the effectiveness of the treatment activity and treatment session and can then modify the treatment as necessary to achieve the stated goals.

Treatment modifications may include:

1. Changing the treatment procedures or the techniques
2. Increasing or decreasing facilitation by changing the use of
   a) reflexes
   b) manual contact
   c) visual cues
   d) verbal cues
   e) traction and approximation

3. Increasing or decreasing the resistance given
4. Working with the patient in positions of function
5. Progressing to using more complex activities

4.6 Treatment Planning

The selection of the most effective treatment depends on the condition of the patient's muscles and joints and any existing medical problems. The therapist should combine and modify the procedures and the techniques to suit the needs of each patient. *The treatment should be intensive, mobilizing the patient's reserves without resulting in pain or fatigue.* The following examples of procedures, techniques, and combinations to treat specific patient problems should not be interpreted as definitive but only as guidelines.

1. Pain
a) Procedures
  I)   Indirect treatment
  II)   Resistance below that which produces pain or stress
  III) Isometric muscle contraction
  IV)   Bilateral work
  V)   Traction
  VI)   Position for comfort

b) Techniques
  I)   Rhythmic stabilization
  II)   Hold-relax
  III) Stabilizing reversals

c) Combinations
  I)   Hold-relax followed by combination of isotonics
  II)   Rhythmic stabilization followed by slow reversal (dynamic reversals) moving first toward the painful range

2. Decreased strength and active range of motion
a) Procedures
  I)   Appropriate resistance
  II)   Timing for emphasis
  III) Stretch
  IV)   Traction or approximation
  V)   Patient position

b) Techniques
  I)   Repeated stretch from beginning of range
  II)   Repeated stretch through range (repeated contractions)
  III) Combination of isotonics
  IV)   Dynamic (slow) reversal of antagonists
        – Facilitation from stronger antagonists
        – Prevention and relief of fatigue

c) Combinations
- I) Dynamic reversal of antagonists combined with repeated stretch through range (repeated contractions) of the weak pattern
- II) Rhythmic stabilization at a strong point in the range of motion followed by repeated contractions of the weak pattern

3. Decreased passive range of motion
a) Procedures
- I) Timing for emphasis
- II) Traction
- III) Appropriate resistance

b) Techniques
- I) Contract-relax or hold-relax
- II) Stabilizing reversal of antagonists
- III) Rhythmic stabilization

c) Combinations
- I) Contract-relax followed by combination of isotonics in the new range
- II) Contract-relax followed by slow Reversals, beginning with motion into the new range
- III) Rhythmic stabilization or stabilizing reversals followed by dynamic reversal of antagonists

4. Coordination and control
a) Procedures
- I) Patterns of facilitation
- II) Manual contact (grip)
- III) Vision
- IV) Proper verbal cues, decreased cuing as patient progresses
- V) Decreasing facilitation as the patient progresses

b) Techniques
- I) Rhythmic initiation
- II) Combination of isotonics
- III) Dynamic reversal of antagonists
- IV) Stabilizing reversals

c) Combinations
- I) Rhythmic initiation, progressing to combination of isotonics
- II) Rhythmic initiation done as reversals, progressing to reversal of antagonists
- III) Combination of isotonics combined with stabilizing or dynamic reversal of antagonists

5. Stability and balance
a) Procedures
- I) Approximation
- II) Vision

III) Manual contact (grip)
IV) Appropriate verbal commands

b) Techniques
    I)   Stabilizing reversals
    II)   Combination of isotonics
    III)   Rhythmic stabilization

c) Combinations
    I)   Dynamic reversal of antagonists progressing to stabilizing reversals
    II)   Dynamic eccentric reversals progressing to stabilizing reversals

6. Endurance

Increasing the patient's general endurance is a part of all treatments. Varying the activity or exercise being done and changing the activity to a different muscle group or part of the body will enable the patient to work longer and harder. Attention to breathing while exercising as well as specific breathing exercises work to increase endurance.

a) Procedure
    I)   Stretch reflex

b) Technique
    I)   Reversal of antagonists

References

Angel RW, Eppler WG Jr (1967) Synergy of contralateral muscles in normal subjects and patients with neurologic disease. Arch Phys Med 48: 233–239

Devine KL, LeVeau BF, Yack J (1981) Electromyographic activity recorded from an unexercised muscle during maximal isomentric exercise of the contralateral agonists and antagonist. Phys Ther 61: 898–903

Hellebrandt FA, Parrish AM, Houtz SJ (1947) Cross education, the influence of unilateral exercise on the contralateral limb. Arch Phys Med 28: 76–85

Markos PD (1979) Ipsilateral and contralateral effects of proprioceptive neuromuscular facilitation techniques on hip motion and electromyographic activity. Phys Ther 59 (11): 1366–1373

Moore JC (1975) Excitation overflow: an electromyographic investigation. Arch Phys Med Rehabil 59: 115–120

Pink M (1981) Contralateral effects of upper extremity proprioceptive neuromuscular facilitation patterns. Phys Ther 61 (8): 1158–1162

Further Readings

Exercise

Engle RP, Canner GG (1989) Proprioceptive neuromuscular facilitation (PNF) and modified procedures for anterior cruciate ligament (ACL) instability. J Orthop Sports Phys Ther 11: 230–236

Hellebrandt FA (1951) Cross education: ipsilateral and contralateral effects of unimanual training. J Appl Physiol 4: 135–144

Hellebrandt FA, Hautz SJ (1950) Influence of bimanual exercise on unilateral work capacity. J Appl Physiol 2: 446–452

Hellebrandt FA Houtz SJ (1958) Methods of muscle training: the influence of pacing. Phys Ther 38: 319–322

Hellebrandt FA, Houtz SJ, Eubank RN (1951) Influence of alternate and reciprocal exercise on work capacity. Arch Phys Med 32: 766–776

Hellebrandt FA, Houtz SJ, Hockman DE, Partridge MJ (1956) Physiological effects of simultaneous static and dynamic exercise. Am J Phys Med 35: 106–117

Nelson AG, Chambers RS, McGown CM, Penrose KW (1986) Proprioceptive neuromuscular facilitation versus weight training for enhancement of muscular strength and athletic performance. J Orthop Sports Phys Ther 8: 250–253

Osternig LR, Robertson RN, Troxel RK, Hansen P (1990) Differential responses to proprioceptive neuromuscular facilitation (PNF) stretch techniques. Med Sci Sports Exerc 22: 106–111

Partridge MJ (1962) Repetitive resistance exercise: a method of indirect muscle training. Phys Ther 42: 233–239

Pink M (1981) Contralateral effects of upper extremity proprioceptive neuromuscular facilitation patterns. Phys Ther 61: 1158–1162

Richardson C, Toppenberg R, Jull G (1990) An initial evaluation of eight abdominal exercises for their ability to provide stabilization for the lumbar spine. Aust Physiother 36: 6–11

Hemiplegia

Brodal A (1973) Self-observations and neuro-anatomical considerations after a stroke. Brain 96: 675–694

Duncan PW, Nelson SG (1983) Weakness – a primary motor deficit in hemiplegia. Neurol Rep 7 (1): 3–4

Harro CC (1985) Implications of motor unit characteristics to speed of movement in hemiplegia. Neurol Rep 9 (3): 55–61

Tang A, Rymer WZ (1981) Abnormal force–EMG realtions in paretic limbs of hemiparentic human subjects. J Neurol Neurosurg Psychiatry 44: 690–698

Trueblood PR, Walker JM, Perry J, Gronley JK (1988) Pelvic exercise and gait in hemiplegia. Phys Ther 69: 32–40

Whitley DA, Sahrmann SA, Norton BJ (1982) Patterns of muscle activity in the hemiplegic upper extremity. Phys Ther 62: 641

Winstein CJ, Jewell MJ, Montgomery J, Perry J, Thomas L (1982) Short leg casts: an adjunct to gait training hemiplegics. Phys Ther 64: 713–714

Motor Control and Motor Learning

APTA (1991) Movement Science: an American Physical Therapy Association monograph. APTA, Alexandria, VA

APTA (1991) Contemporary management of motor control problems. Proceedings of the II Step conference. Foundation for Physical Therapy, Alexandria, VA

Hellebrandt FA (1958) Application of the overload principle to muscle training in man. Arch Phys Med Rehabil 37: 278–283

Light KE (1990) Information processing for motor performance in aging adults. Phys Ther 70 (12): 820–826

VanSant AF (1988) Rising from a supine position to erect stance: description of adult movement and a developmental hypothesis. Phys Ther 68 (2): 185–192

VanSant AF (1990) Life-span development in functional tasks. Phys Ther 70 (12): 788–798

Spasticity

Landau WM (1974) Spasticity: the fable of a neurological demon and the emperor's new therapy. Arch Neurol 31: 217–219

Levine MG, Kabat H, Knott M, Voss DE (1954) Relaxation of spasticity by physiological technics. Arch Phys Med Rehabil 35: 214–223

Perry J (1980) Rehabilitation of spasticity. In: Felman RG, Young JRR, Koella WP (eds) Spasticity: disordered motor control. Year Book, Chicago

Sahrmann SA, Norton BJ (1977) The relationship of voluntary movement to spasticity in the upper motor neuron syndrome. Annal Neurol 2: 460–465

Young RR Wiegner AW (1987) Spasticity. Clin Orthop 219: 50–62

Cold

Baker RJ, Bell GW (1991) The effect of therapeutic modalities on blood flow in the human calf. J Orthop Sports Phys Ther 13: 23–27

Miglietta O (1964) Electromyographic characteristics of clonus and influence of cold. Arch Phys Med Rehabil 45: 508–512

Miglietta O (1962) Evaluation of cold in spasticity. Am J Phys Med 41: 148–151

Olson JE, Stravino VD (1972) A review of cryotherapy. Phys Ther 52: 840–853

Prentice WE Jr (1982) An electromyographic analysis of the effectiveness of heat or cold and stretching for inducing relaxation in injured muscle. J Orthop Sports Phys Ther 3: 133–140

Sabbahi MA, Powers WR (1981) Topical anesthesia: a possible treatment method for spasticity. Arch Phys Med Rehabil 62: 310–314

5 Patterns of Facilitation

Normal functional motion is composed of mass movement patterns of the limbs and the synergistic trunk muscles (Kabat 1960). The motor cortex generates and organizes these movement patterns, and the individual cannot, voluntarily, leave a muscle out of the movement pattern to which it belongs. This does not mean that we cannot contract muscles individually, but our discrete motions spring from the mass patterns (Beevor 1978; Kabat 1950). These synergistic muscle combinations form the PNF patterns of facilitation.

Some people believe that you must know and use the PNF patterns to work within the concept of PNF. We think that you need only the philosophy and the appropriate procedures. The patterns, while not essential, are however a valuable tool to have in the armamentarium. Working with the synergistic relationships in the patterns allows problems to be treated indirectly. The stretch reflex is more effective when an entire pattern rather than just the individual muscle is stretched.

The PNF patterns combine motion in all three planes:

1. The sagittal plane: flexion and extension
2. The coronal or frontal plane: abduction and adduction of limbs or lateral flexion of the spine
3. The transverse plane: rotation

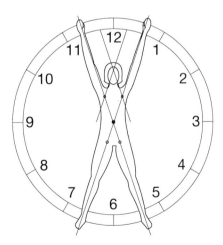

Fig. 5.1. Patterns are "spiral and diagonal" (modified, from Klein-Vogelbach, 1990: Functional Kinetics. Springer Berlin Heidelberg New York)

We thus have motion that is "spiral and diagonal" (Knott and Voss 1968) (Fig. 5.1). Stretch and resistance reinforce the effectiveness of the patterns, as shown by an increased activity in the muscles. The increased muscular activity spreads both distally and proximally within a pattern and from one pattern to related patterns of motion (irradiation). Treatment makes use of irradiation from those synergistic combinations of muscles (patterns) to strengthen the desired muscle groups or reinforce the desired functional motions.

When we exercise in the patterns against resistance, all the muscles that are a part of the synergy will contract if they can. The rotational component of the pattern is the key to effective resistance. Correct resistance to rotation will strengthen the entire pattern. Too much resistance to rotation will prevent motion from occurring or "break" a stabilizing contraction.

The motion occurring at the *proximal* joint names the patterns, as in flexion–adduction–external rotation of the shoulder. Two antagonistic patterns make up a *diagonal,* for example, shoulder flexion–adduction–external rotation and the antagonist pattern extension–abduction–internal rotation.

The proximal and distal joints of the limb are linked in the pattern. The middle joint is free to flex, extend or maintain its position. For example, finger flexion, radial flexion of the wrist, and forearm supination are integral parts of the pattern of shoulder flexion–adduction–external rotation. The elbow, however, may flex, extend or remain in one position.

The trunk and limbs work together to form complete synergies. For example, the pattern of shoulder flexion–adduction–external rotation with anterior elevation of the scapula combines with trunk extension and rotation to the opposite side to complete a total motion. If you know the synergistic muscle combinations, you can work out the patterns. If you know the pattern, you will know the synergistic muscles.

The *groove* of the pattern is that line drawn by the hand or foot (distal component) as the limb moves through its range. For the *head and neck,* the groove is

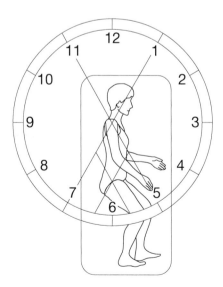

Fig. 5.2. The groove

drawn by a plane through the nose, chin, and crown of the head. The groove for the *upper trunk* is drawn by the tip of the shoulder and for the *lower trunk* by the hip. Because the trunk and limbs work together, their grooves join or are parallel (Fig. 5.2). As discussed earlier, the therapist's body should be in line with or parallel to the relevant groove. Pictures of the complete patterns with the therapist in the proper position come in the following chapters.

To move concentrically through the entire range of a pattern:

1. The limb is positioned in the "lengthened range."
   a) All the associated muscles (agonists) are lengthened.
   b) There is *no pain,* and no joint stress.
   c) The trunk does not rotate or roll.
2. The limb moves into the "shortened range."
   a) The end of the range of contraction of the muscles (agonists) is reached.
   b) The antagonistic muscle groups are lengthened.
   c) There is *no pain* and no joint stress.
   d) The trunk did not rotate or roll.

The normal timing of the pattern is:

1. The distal part (hand and wrist or foot and ankle) moves through its full range first and holds its position.
2. The other components move smoothly together so that they complete their movement almost simultaneously.
3. Rotation is an integral part of the motion and is resisted from the beginning to the end of the motion.

We can vary the pattern in several ways:

1. By changing the activity of the middle joint in the extremity pattern for function.
   Example: First, the pattern of shoulder flexion–abduction–external rotation is done with the elbow moving from extension to flexion. The patient's hand rubs his or her head. The next time, the same pattern is done with the elbow moving from a flexed to an extended position, so the patient's hand can reach for a high object.
2. By changing the activity of the middle joint in the extremity pattern for the effect on two-joint muscles.
   Example: First, the pattern of hip flexion–adduction– external rotation is done with the knee moving from the extended to the flexed position. In this combination, the hamstring muscles shorten actively. Next time, the same pattern is used with the knee staying straight. This combination stretches the hamstring muscles.
3. By changing the patient's position to change the effects of gravity.
   Example: The pattern of hip extension–abduction–internal rotation is done in a sidelying position so the abductor muscles are working against gravity.
4. By changing the patient's position to a more functional one.
   Example: The arm patterns are exercised in a sitting position and functional activities such as eating incorporated.
5. By changing the patient's position to use visual cues.

Example: Have the patient in a half-sitting position so that he or she can see when exercising the foot and ankle.

We can combine the patterns in many ways. The emphasis of treatment is on the arms or legs when the limbs move independently. The emphasis is on the trunk when the arms are joined by one hand gripping the other arm or the legs touch and move together. Choosing how to combine the patterns for the greatest functional effect is a part of the assessment and treatment planning.

We name the pattern combinations according to how the movement of the limbs (arms, legs or both) relate to each other:

1. Unilateral: one arm or one leg
2. Bilateral: both arms, both legs, or combinations of arms and legs:
   a) symmetrical: both move, in the same pattern (e. g., both flexion–adduction)
   b) asymmetrical: both move in opposite patterns (e. g., right, flexion–adduction; left, flexion–abduction)
   c) symmetrical reciprocal: both move in the same diagonal but opposite directions (e. g., right, flexion–adduction; left, extension–abduction)
   d) asymmetrical reciprocal: both move in opposite diagonals and opposite directions (e. g., right, flexion–adduction; left, extension–adduction)

References

Beevor CE (1978) The Croonian lectures on muscular movements and their representation in the central nervous system. In: Payton OD, Hirt S, Newton RA (eds) Scientific basis for neurophysiologic approaches to therapeutic exercise: an anthology. Philadelphia Davis, Philadelphia

Kabat H (1950) Studies on neuromuscular dysfunction, XIII: New concepts and techniques of neuromuscular reeducation for paralysis. Perm Found Med Bull 8 (3): 121–143

Kabat H (1960) Central mechanisms for recovery of neuromuscular function. Science 112: 23–24

Knott M, Voss DE (1968) Proprioceptive neuromuscular facilitation: patterns and techniques, 2nd edn. Harper and Row, New York

Further Reading

Bosma JF, Gellhorn E (1946) Electromyographic studies of muscular co-ordination on stimulation of motor cortex. J Neurophysiol 9: 263–274

Gellhorn E (1948) The influence of alterations in posture of the limbs on cortically induced movements. Brain 71: 26–33

6 The Scapula and Pelvis

6.1 Introduction

Exercise of the scapula and pelvis is important for treatment of the neck, the trunk, and the extremities. Although the scapula is not directly attached to the spine, the scapular muscles control or influence the function of the cervical and thoracic spine. Proper function of the upper extremities requires both motion and stability of the scapula. Pelvic motion and stability are required for proper function of the trunk and the lower extremities.

6.1.1 Applications

Exercise of the scapula and pelvis can have various goals:

1. Scapula
   a) Exercise the scapula independently for motion and stability.
   b) Exercise trunk muscles.
      I)   Using timing for emphasis, prevent scapular motion at the beginning of the range until you feel and see the trunk muscles contract. When this occurs, change the resistance at the scapula so that both the scapula and the trunk motion are resisted.
      II)   At the end of the scapular range of motion, "lock in" the scapula with a stabilizing contraction and exercise the trunk with repeated contractions.
      III)   Use reversal of antagonist techniques to train coordination and prevent or reduce fatigue of the scapular and trunk muscles.
   c) Exercise functional activities.
      I)   When the trunk muscles are contracting you can extend their action into such functional activities as rolling forward or backward (see Sect. 11.3.1). Give a movement command such as "roll forward" and resist the functional activity using the stabilized scapula as the handle.
      II)   Repeated contractions of the functional activity will reinforce both learning the activity and the physical ability to perform it.
   d) Exercise the neck.
      I)   Resist the moving or stabilizing contraction at the scapula and head simultaneously to exercise the muscles that go from the cervical spine to the scapula.
      II)   To stretch these muscles, stabilize the cervical spine and resist the appropriate scapular motion.

e) Facilitate arm motion and stability (by resisting scapular motion and stabilization, since the scapula and arm muscles reinforce each other).
  I)   Scapular elevation patterns work with arm flexion patterns.
  II) Scapular depression patterns work with arm extension patterns.

2. Pelvis
  a) Exercise trunk muscles.
    I)   Resist the pelvic patterns to exercise lower trunk flexor, extensor, and lateral flexor muscles. The pelvis should not move further into an anterior or posterior tilt during these exercise.
    II) Use repeated stretch from beginning of range or through range to strengthen these trunk muscles.
    III) Use reversal of antagonist techniques to train coordination and prevent or reduce fatigue of the working muscles.
  b) Exercise functional trunk activities.
    I)   Use a stabilizing contraction to lock in the pelvis, then give a functional command such as "roll" and resist the activity using the stabilized pelvis as the handle (see Sect. 11.3.1).
    II) Use repeated contractions to strengthen and reinforce learning of the functional activity.
    III) Use the technique combination of isotonics to teach control of the trunk motions. Have the patient control trunk motion with concentric and eccentric contractions while maintaining the pelvic stabilization.
    IV) Use reversal techniques to prevent or relieve muscular fatigue.
  c) Treat the upper trunk and cervical areas indirectly through irradiation. Give sustained maximal resistance to stabilizing or isometric pelvic patterns until you see and feel contraction of the desired upper trunk and cervical muscles.
  d) The pelvis and lower extremities reinforce each other.
    I)   Pelvic depression patterns work with weight-bearing motions of the legs.
    II) Pelvic elevation patterns work with stepping or leg lifting motions.

6.1.2 Diagonal Motion

The scapular and pelvic patterns occur in two diagonals: anterior elevation – posterior depression and posterior elevation – anterior depression. Movement in the diagonals is an arc that follows the curve of the patient's torso. When the scapula or pelvis is moved within the diagonal, the patient will not roll forward or back or rotate around one spinal segment.

Picture a patient lying on the left side (Fig. 6.1). Now imagine a clock with the 12 o'clock position toward the patient's head, the 6 o'clock position toward the feet, the 3 o'clock position anterior and the 9 o'clock position posterior. When working with the right scapula or pelvis, anterior elevation is toward 1 o'clock and posterior depression toward 7 o'clock. Posterior elevation is toward 11 o'clock and anterior depression toward 5 o'clock (Fig. 6.1).

Now imagine that the patient is lying on the right side. The 12 o'clock position is still towards the patient's head but the 3 o'clock position is posterior and the 9 o'clock position anterior. Working with the left scapula or pelvis, anterior ele-

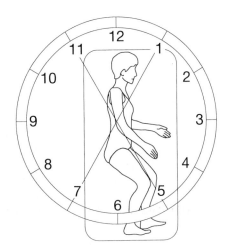

Fig. 6.1. Diagonal motion

vation is toward 11 o'clock and posterior depression toward 5 o'clock; posterior –
elevation is toward 1 o'clock and anterior depression toward 7 o'clock. In this
chapter we show all patterns being done on the patient's left scapula or pelvis. All
references are to motion of the left scapula or the left side of the pelvis.

6.1.3 Patient Position

The procedures start with the patient in a stable sidelying position, the hips and
knees flexed to 90°. The patient should be positioned that so his or her back is close
to the edge of the treatment table. The patient's spine is maintained in a normal
alignment and the head and neck in as neutral a position as possible, neither flexed
nor extended. The patient's head is supported in line with the spine, avoiding lateral bend.

   Before beginning a scapula or pelvis pattern, place the scapula or pelvis in a mid-
position where the line of the two diagonals cross. The scapula should not be rotated, and the glenohumeral complex should lie in the anteroposterior midline. The
pelvis should be in the middle, between anterior and posterior tilt. From this midline position, the scapula or pelvis can then be moved into the elongated range of
the pattern.

6.1.4 Therapist Position

The therapist stands behind the patient, facing the line of the scapular or pelvic diagonal and with arms and hand aligned with the motion. All the grips described in
this chapter assume that the therapist is in this position.

   In an alternative position the patient lies facing the edge of the treatment table.
The therapist stands in front of the patient in the line of the chosen diagonal. The
hand placement on the patient's body remains the same but the grips use different
areas of the therapist's hands (Fig. 6.2).

49

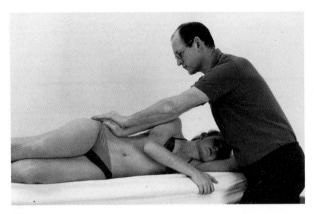

Fig. 6.2. The therapist is in front of the patient: anterior elevation of the pelvis

The scapular and pelvic patterns can also be done with the patient lying on the mats. In this position, the therapist must kneel on the mats either in front of or behind the patient. Weight shifting is done by moving from the position of sitting on the heels (kneeling down) to partial or fully upright kneeling (kneeling up).

6.1.5 Grips

The grips follow the basic procedure for manual contact, that is opposite the direction of movement. This section describes the two-handed grips used when the patient is sidelying and the therapist is standing behind the patient. These grips are modified when the therapist's or patient's position is changed, and some modification is also needed when the therapist can use only one hand while the other hand controls another pattern or extremity.

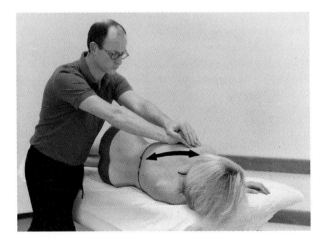

Fig. 6.3 a

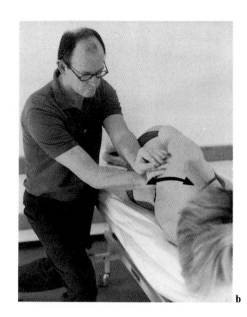

Fig. 6.3 a, b. The scapula and the direction of the therapist's resistance move in an arc

b

6.1.6 Resistance

The direction of the resistance is an arc following the contour of the patient's body. The angle of the therapist's hands and arms changes as the scapula or pelvis moves through this arc of diagonal motion (Fig. 6.3).

6.2 Scapula Patterns

Scapula patterns can be done with the patient lying on the treatment table, on mats, sitting, or standing. Sidelying (illustrated) allows free motion of the scapula and easy reinforcement of trunk activities. The main muscle components are as follows (extrapolated from Kendall and McCreary 1983). We know of no confirming electromyographic studies.

Movement	Muscles: principal components
Anterior elevation	Levator scapulae, rhomboids, serratus anterior
Posterior depression	Serratus anterior (lower), rhomboids, latissimus dorsi
Posterior elevation	Trapezius, levator scapulae
Anterior Depression	Rhomboids, serratus anterior, pectoralis minor and major

6.2.1 Anterior Elevation and Posterior Depression

Standing behind the patient you face up and forward.

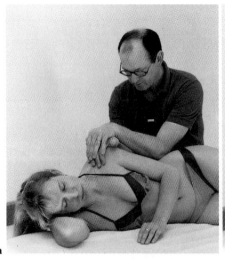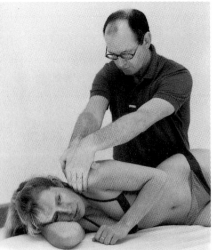

a b

Fig. 6.4 a, b. Scapula: anterior elevation

6.2.1.1 Anterior Elevation (Fig. 6.4)

Grip. Place one hand on the anterior aspect of the glenohumeral joint and the acromion with your fingers cupped. The other hand covers and supports the first. Contact is with the fingers and not the palm of the hand.

Elongated Position. Pull the entire scapula down and back toward the lower thoracic spine (posterior depression). Be sure that the glenohumeral complex is positioned posterior to the central anteroposterior line of the body (mid-frontal plane). You should see and feel that the anterior muscles of the neck are taut. Do not pull so far that you lift the patient's head up. Continued pressure on the scapula should not cause the patient to roll back or rotate the spine around one segment.

Command. "Shrug your shoulder up toward your nose." "Pull".

Movement. The scapula moves up and forward in a line aimed approximately at the patient's nose.

Body Mechanics. Keep your arms relaxed and let your body give the resistance by shifting your weight from the back to the front leg.

Resistance. The line of resistance is an arc following the curve of the patient's body. Start with your elbows low and your forearms parallel to the patient's back. At the end of the pattern your elbows are extending and you are lifting upward.

End Position. The scapula is up and forward with the acromion close to the patient's nose. The scapular retractor and depressor muscles are taut.

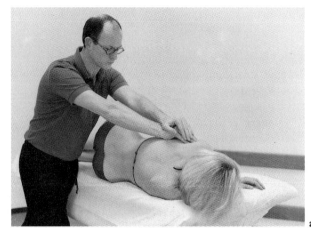

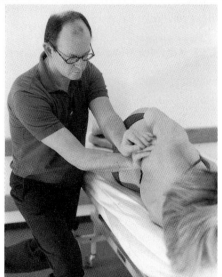

Fig. 6.5 a, b. Scapula: posterior depression

6.2.1.2 Posterior Depression (Figs. 6.3, 6.5)

Grip. Place the heels of your hands along the vertebral border of the scapula with one hand just above (cranial to) the other. Your fingers lie on the scapula pointing toward the acromion. Try to keep all pressure below (caudal to) the spine of the scapula.

Elongated Position. Push the scapula up and forward (Anterior elevation) until you feel and see that the posterior muscles below the spine of the scapula are tight. Continued pressure should not cause the patient to roll forward or rotate the spine around one segment.

Command. "Push your shoulder blade down to me." "Push".

Movement. The scapula moves down (caudal) and back (adduction), toward the lower thoracic spine.

Body Mechanics. Flex your elbows to keep your forearms parallel to the line of resistance. Shift your weight to your back foot and allow your elbows to drop as the patient's scapula moves down and back (Fig. 6.5).

Resistance. The line of resistance is an arc following the curve of the patient's body. Start by lifting the scapula down towards the patient's nose. As the scapula moves toward the anteroposterior midline the resistance is forward and almost parallel to the supporting table. By the end of the motion the resistance is forward and upward toward the ceiling.

End Position. The scapula is depressed and retracted with the glenohumeral complex posterior to the central anteroposterior line of the trunk. The vertebral border should lie flat and not wing out.

6.2.2 Anterior Depression and Posterior Elevation

Standing behind the patient's head you face down towards the patient's bottom (right) hip.

6.2.2.1 Anterior Depression (Fig. 6.6)

Grip. Place one hand posteriorly with the fingers holding the lateral (axillary) border of the scapula. The other hand holds anteriorly on the axillary border of the pectoralis major muscle and the coracoid process. The fingers of both hands point toward the opposite ilium, and your arms are lined up in the same direction.

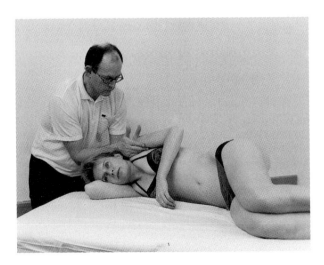

Fig. 6.6 a

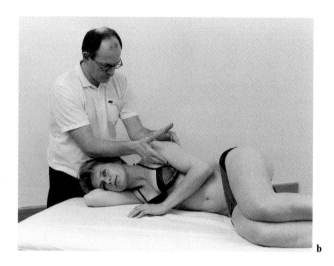

Fig. 6.6a, b. Scapula:
an terior depression

Elongated Position. Lift the entire scapula up and back toward the middle of the back of the head (posterior elevation). Be sure that the glenohumeral complex is positioned posterior to the central anteroposterior line of the body (mid-frontal plane). You should see and feel that the abdominal area is taut from the ipsilateral ribs to the contralateral pelvis. Continued pressure on the scapula should not cause the patient to roll back or rotate the spine around one segment.

Command. "Pull you shoulder blade down towards your navel." "Pull."

Movement. The scapula moves down and forward, in a line aimed at the opposite anterior iliac crest.

Body Mechanics. Let the resistance come from your body weight as you shift from the back to the front leg.

Resistance. The resistance follows the curve of the patient's body. At the end of the pattern you are lifting in a line parallel to the front of the patient's thorax.

End Position. The scapula is rotated forward, depressed, and abducted. The glenohumeral complex is anterior to the central anteroposterior line of the body.

6.2.2.2 Posterior Elevation (Fig. 6.7)

Grip. Place your hands posterior on the upper trapezius muscle, staying above (superior to) the spine of the scapula. Overlap your hands as necessary to stay distal to the junction of the spine and first rib.

Elongated Position. Round the scapula down and forward toward the opposite ilium (anterior depression) until you feel that the upper trapezius muscle is taut. Do not push so far that you lift the patient's head up. Continued pressure should not cause the patient to roll forward or rotate the spine around one segment.

55

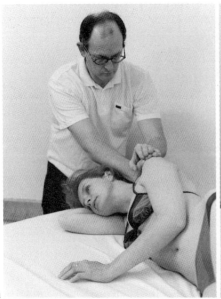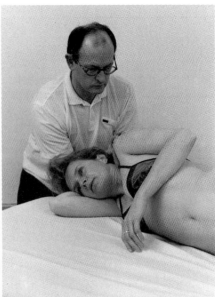

Fig. 6.7 a, b. Scapula: posterior elevation

Command. "Shrug your shoulder up." "Push".

Movement. The scapula shrugs up (cranially) and back (adduction) in a line aimed at the middle of the top of the patient's head. The glenohumeral complex moves posteriorly and rotates upward.

Body Mechanics. Shift your weight from the front to the back foot as the scapula moves. Your forearms stay parallel to the line of resistance.

Resistance. The resistance follows the curve of the patient's body. At the end of the pattern you are lifting around the thorax and away from top of the patient's head.

End Position. The scapula is elevated and adducted, the glenohumeral complex is posterior to the central anteroposterior line of the body.

6.3 Pelvis Patterns

Pelvis patterns can be done with the patient lying, sitting, or standing. The side that is moving must not be weight-bearing. The pelvis is part of the trunk, so the range of motion in the pelvic patterns depends on the amount of motion in the lower spine. We treat pelvic patterns as isolated from the trunk if no increased lumbar flexion or extension occurs. Sidelying (illustrated) allows free motion of the pelvis and easy reinforcement of trunk and lower extremity activities. The movements and muscle components mainly involved are as follows (Kendall and McCready 1983):

Movement	Muscles: principal components
Anterior elevation	Internal and external oblique abdominal muscles
Posterior depression	Contralateral quadratus lumborum, iliocostalis lumborum, and longissimus thoracis
Posterior elevation	Ipsilateral quadratus lumborum, ipsilateral latissimus dorsi, iliocostalis lumborum, and longissimus thoracis
Anterior depression	Contralateral internal and external oblique abdominal muscles

6.3.1 Anterior Elevation and Posterior Depression

Standing behind the patient your body faces up and towards the patient's lower (right) shoulder.

6.3.1.1 Anterior Elevation (Fig. 6.8)

Grip. The fingers of one hand grip around the crest of the ilium, on and just anterior to the midline. Your other hand overlaps the first.

Elongated Position. Pull the crest of the pelvis back and down in the direction of posterior depression. See and feel that the tissues stretching from the crest of the ilium to the opposite rib cage are taut. Continued pressure should not cause the patient to roll backward or rotate the spine around one segment.

Command. "Shrug your pelvis up." "Pull."

Movement. The pelvis moves up and forward without tilt. There is an anterior shortening of the trunk on that side (lateral flexion).

Body Mechanics. Start with your elbows flexed to pull the iliac crest down as well as back. As the movement progresses your elbows extend and your weight shifts from your back to your front foot.

Resistance. The line of resistance curves following the patient's body. Start by pulling the pelvis back towards you and down towards the table. As the pelvis moves to the mid-position the line of the resistance is almost straight back. At the end of the motion the resistance is up towards the ceiling.

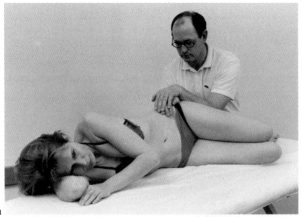

a

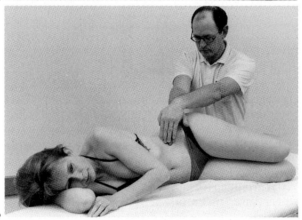

b

Fig. 6.8 a, b. Pelvis: anterior elevation

End Position. The pelvis is up (elevated) and forward (anterior) toward the lower shoulder with no increase in the anterior or posterior tilt. The upper side (left) of the trunk is shortened and laterally flexed with no change in lumbar lordosis.

6.3.1.2 Posterior Depression (Fig. 6.9)

Grip. Place the heel of one hand on the ischial tuberosity. Overlap and reinforce the hold with your other hand. The fingers of both hands point diagonally forward.

Elongated Position. Push the ischial tuberosity up and forward to bring the iliac crest down closer to the opposite rib cage (anterior elevation). Continued pressure should not cause the patient to roll forward or rotate the spine around one segment.

Command. "Sit into my hand." "Push."

Movement. The pelvis moves down and posteriorly without tilt. There is an elongation of the trunk on that side without an increase in lumbar lordosis.

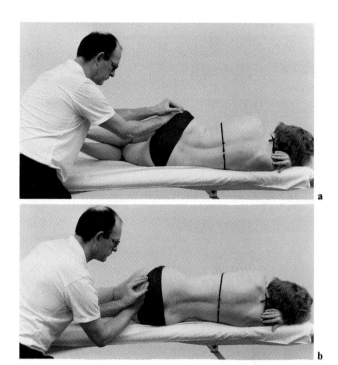

Fig. 6.9 a, b. Pelvis: posterior depression

Body Mechanics. Your elbows flex as the patient's pelvis moves downward and you shift your weight from your front to your back foot.

Resistance. The resistance is always upward on the ischial tuberosity while pushing diagonally forward (anterior and cranial).

End Position. The pelvis is down and back (posterior) with no increase in the anterior or posterior tilt. The upper side (left) of the trunk is elongated with no increase in lumbar lordosis.

6.3.2 Anterior Depression and Posterior Elevation

Standing behind the patient, face toward a line representing about 25° of flexion of the patient's bottom (right) leg.

6.3.2.1 Anterior Depression

Grip. Place the fingers of one hand on the greater trochanter of the femur. The other hand may reinforce the first hand (Fig. 6.10a) or you may grip below the anterior inferior iliac spine.

Alternate Grip. The fingers of the posterior hand grip the ischial tuberosity. The anterior hand grips below the anterior inferior iliac spine.

59

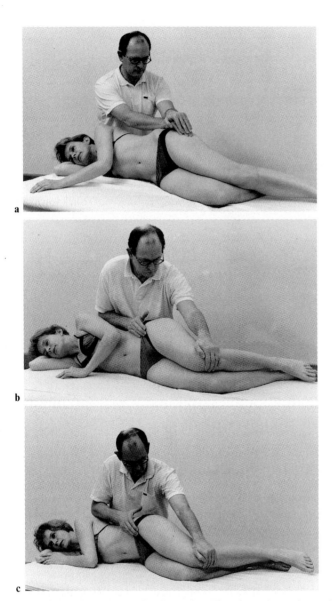

a

b

c

Fig. 6.10 a–c. Pelvis: anterior depression. The grip on the trochanter is shown in *a*

For a grip using the leg place your right hand on the patient's anteroinferior iliac spine and your left hand on the patient's left knee (Fig. 6.10b, c) You must move the patient's leg until the femur is in the line of the pattern (about 20°–30° of hip flexion) (Fig. 6.10b).

Elongated Position. Gently move the pelvis up (cranial) and back (dorsal) towards the lower thoracic spine (posterior elevation). Be careful not to rotate or compress spinal joints.

Command. "Pull down and forward." ("Push your knee into my hand.")

Movement. The pelvis moves down and anteriorly without tilt. There is an elongation of the trunk on that side without an increase in lumbar lordosis.

Body Mechanics. Start with your elbows flexed to keep your forearms parallel to the patients back. Shift your weight to your front foot during the motion and allow your elbows to extend.

Resistance. At the beginning of the movement the resistance is toward the patient's lower thoracic spine. As the motion continues, the line of the resistance follows the curve of the body. At the end of the pattern the resistance is diagonally back toward you and up toward the ceiling.

End Position. The pelvis is down and forward with no increase in the anterior or posterior tilt. The trunk is elongated with no increase in lumbar lordosis.

6.3.2.2 Posterior Elevation (Fig. 6.11)

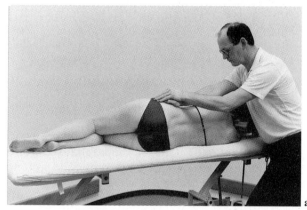

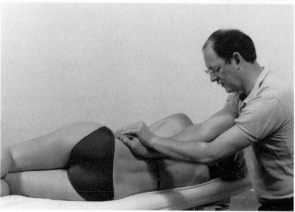

Fig. 6.11 a, b. Pelvis: posterior elevation

61

Grip. Put the heel of one hand on the crest of the ilium, on and just posterior to the midline. Your other hand overlaps the first. There is no finger contact.

Elongated Position. Gently push the pelvis down and forward until you feel and see that the posterior lateral tissues on that side are taut (anterior depression). Continued pressure should not cause the patient to roll forward or rotate the spine around one segment.

Command. "Push your pelvis up and back – gently."

Movement. The pelvis moves up and back without tilt. There is a posterior shortening of the trunk on that side (lateral flexion).

Body Mechanics. As the pelvis moves up and back shift your weight to your back foot. At the same time flex and drop your elbows so that they point down towards the table.

Resistance. The resistance begins by lifting the posterior iliac crest around toward the front of the table. At the end of the motion the resistance has made an arc around the body and is now lifting the ilia crest up toward the ceiling.

End Position. The pelvis is up and back with no increase in the anterior or posterior tilt. The upper side (left) of the trunk is shortened and laterally flexed with no increase in lumbar lordosis.

6.4 Symmetrical, Reciprocal and Asymmetrical Exercises

In addition to the exercises carried out with one body part in one direction (the scapula moving into anterior elevation) and in both directions (the scapula moving back and forth between anterior elevation and posterior depression), both scapulae or the scapula and pelvis can be exercised simultaneously. Any combination of scapular and pelvic patterns may be used, limited only by the patient's abilities and

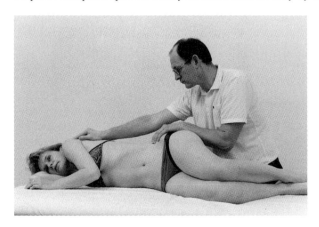

Fig. 6.12 a

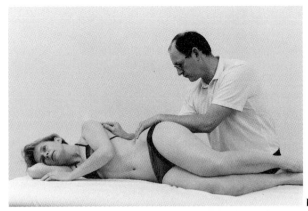

Fig. 6.12 a, b. Symmetrical-reciprocal exercise: the scapula moves in anterior elevation, the pelvis in posterior depression

the therapist's imagination. Here we describe and illustrate two combinations. Use the basic procedures (grip, command, resistance, timing, etc.) and techniques when you use the symmetrical and asymmetrical pattern combinations just as when you work with single patterns in one direction.

6.4.1 Symmetrical-Reciprocal Exercise:

Anterior Elevation – Posterior Depression of the Scapula and Pelvis

Here the scapula and pelvis move in the same diagonal but in opposite patterns (Figs. 6.12, 6.13). You position your body parallel to the lines of the diagonals.

This combination of scapular and pelvic motions results in full trunk elongation and shortening with counterrotation. The motion is an enlarged version of the normal motion of scapula, pelvis and trunk during walking.

Pelvis	*Scapula*
Anterior elevation	Posterior depression (Fig. 6.12)
Posterior depression	Anterior elevation (Fig. 6.13)

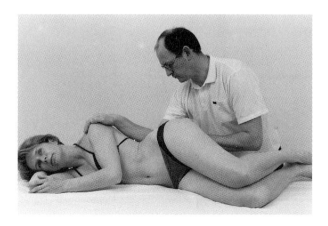

Fig. 6.13 a

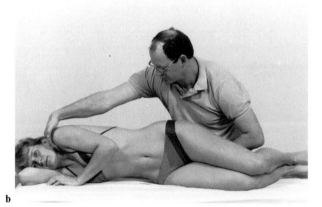

Fig. 6.13 a, b. Symmetrical-reciprocal exercise: the scapula moves in anterior elevation, the pelvis in posterior depression

6.4.2 Asymmetrical Exercise:

Anterior Elevation – Posterior Depression of the Pelvis with Anterior Depression – Posterior Elevation of the Scapula

In this combination the scapula and pelvis move in opposite diagonals and the diagonals are not parallel (Figs. 6.14, 6.15). Position your body in the middle and align your forearms so that one is in the line of each diagonal. You cannot use your body weight for resistance with this combination.

When both the scapula and pelvis move in the anterior patterns (forward toward each other) the result is mass trunk flexion. When they both move in the posterior patterns (backward away from each other) the result is mass trunk extension with elongation.

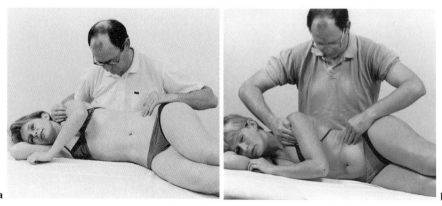

Fig. 6.14 a, b. Asymmetrical exercise: the scapula moves in anterior depression, the pelvis in anterior elevation

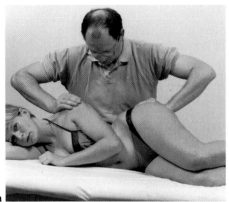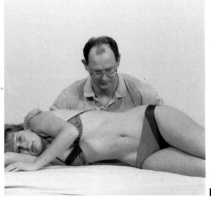

Fig. 6.15 a, b. Asymmetrical exercise: the scapula moves in posterior elevation, the pelvis in posterior depression

Pelvis   *Scapula*
Anterior elevation   Anterior depression (Fig. 6.14)
Posterior depression   Posterior elevation (Fig. 6.15)

Reference

Kendall FP, McCreary EK (1983) Muscles, testing and function. Wiliams and Wilkins, Baltimore

7 The Upper Extremity

7.1 Arm Patterns

7.1.1 Introduction

Upper extremity patterns are used to treat dysfunction caused by muscular weakness, incoordination, and joint restrictions. The arm patterns are also used to exercise the trunk. Resistance to strong arm muscles produces irradiation to weaker muscles elsewhere in the body.

7.1.1.1 Diagonal Motion

The arm has two diagonals:

1. Flexion-abduction-external rotation and extension-adduction-internal rotation
2. Flexion-adduction-external rotation and extension-abduction-internal rotation

Scapular motion is an integral part of each pattern.

The basic patterns of the left arm with the subject supine are shown in Fig. 7.1. All descriptions refer to this arrangement. To work with the right arm just change the word "left" to "right" in the instructions. Variations of position are shown later in the chapter.

7.1.1.2 Patient Position

Position the patient close to the left edge of the table. Support the patient's head and neck in a comfortable position, as close to neutral as possible. Before beginning an upper extremity pattern, place the patient's arm in a middle position where the lines of the two diagonals cross. The shoulder and forearm should be in neutral rotation. From this midline position, move the extremity into the elongated range of the pattern.

7.1.1.3 Therapist Position

The therapist stands on the left side of the table facing the line of the diagonal, arms and hands aligned with the motion. All grips described in the first part of this chapter (Sect. 7.1) assume that the therapist in this position.

We give the basic position and body mechanics for exercising the straight arm pattern. When we describe variations in the patterns we identify any changes in po-

Shoulder: Flexion adduction
external rotation

Scapula: anterior elevation
Forearm: supination
Wrist: radial flexion
Fingers: radial flexion
Thumb: flexion, adduction

Shoulder: Flexion abduction
external rotation

Scapula: posterior elevation
Forearm: supination
Wrist: radial extension
Fingers: radial extension
Thumb: extension, abduction

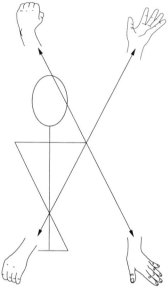

Shoulder: Extension adduction
internal rotation

Scapula: anterior depression
Forearm: pronation
Wrist: ulnar flexion
Fingers: ulnar flexion
Thumb: flexion, opposition

Shoulder: Extension abduction
internal rotation

Scapula: posterior depression
Forearm: pronation
Wrist: ulnar extension
Fingers: ulnar extension
Thumb: extension,
palmar abduction

Fig. 7.1. Upper extremity diagonals (Courtesy of V. Jung)

sition or body mechanics. The therapist's position can vary within the guidelines for the basic procedures. Some of these variations are illustrated at the end of the chapter.

7.1.1.4 Grips

The grips follow the basic procedures for manual contact, that is opposite the direction of movement. The first part of this chapter (Sect. 7.1) describes the two-handed grip used when the therapist stands next to the moving upper extremity. The basic grip is described for each straight arm pattern. The grips are modified when the therapist's or patient's position is changed. The grips also change when the therapist can use only one hand while the other hand controls another pattern or extremity. The grip on the hand contacts the active surface, dorsal or palmar, and holds the sides of the hand to resist the rotary components. Using the lumbrical grip will prevent squeezing or pinching the patient's hand. Remember, pain inhibits effective motion.

7.1.1.5 Resistance

The direction of the resistance is an arc back toward the starting position. The angle of the therapist's hands and arms changes as the limb moves through the pattern. Traction or approximation is an important element of the resistance.

7.1.1.6 Normal Timing

The arm moves through the diagonals in a straight line with rotation occurring smoothly throughout the motion. The hand and wrist (distal component) begin the pattern, moving through their full range. Rotation at the shoulder and forearm accompanies the rotation (radial or ulnar deviation) of the wrist. After the distal movement is completed, the scapula moves with the shoulder or shoulder and elbow jointly through their range.

In the sections on *timing for emphasis* in the patterns described below we offer some suggestions for exercising components of the patterns. Any of the techniques may be used. We have found that repeated stretch (repeated contractions) and combination of isotonics work well, but do not limit yourself to the exercises we suggest: use your imagination.

7.1.2 Flexion-Abduction-External Rotation (Fig. 7.2)

Joint	Movement	Muscles: principal components (Kendall and McCreary 1983)
Scapula	Posterior elevation	Trapezius, levator scapulae, serratus anterior
Shoulder	Flexion, abduction, external rotation	Deltoid (anterior), biceps (long head), coracobrachialis, supraspinatus, infraspinatus, teres minor
Elbow	Extended (position unchanged)	Triceps, anconeus
Forearm	Supination	Biceps, brachioradialis, supinator
Wrist	Radial extension	Extensor carpi radialis (longus and brevis)
Fingers	Extension, radial deviation	Extensor digitorum longus, interossei
Thumb	Extension, abduction	Extensor pollicis (longus and brevis), abductor pollicis longus

Grip. *Distal hand:* Your right hand grips the dorsal surface of the patient's hand. Your fingers are on the radial side (1st and 2nd metacarpal), your thumb gives counterpressure on the ulnar border (5th metacarpal). There is no contact on the palm.

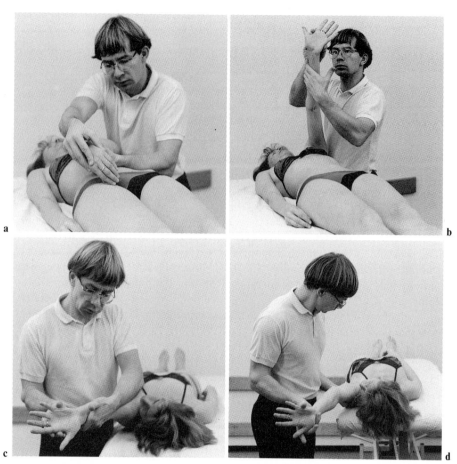

a
b
c
d

Fig. 7.2 a–d. Flexion – abduction – external rotation of the arm. *a* Starting position; *b* middle position; *c* end position; *d* emphasizing the motion of the shoulder

Caution: Do not squeeze the hand.

Proximal hand: From underneath the arm, hold the radial and ulnar sides of the patient's forearm proximal to the wrist. The lumbrical grip allows you to avoid placing any pressure on the anterior (palmar) surface of the forearm.

Alternative Grip. To emphasize shoulder or scapula motions, move the proximal hand to grip the upper arm or the scapula after the wrist completes its motion (Fig. 7.2 d).

Elongated Position. Traction the entire arm and scapula while you move the wrist into ulnar flexion and the forearm into pronation. Continue the traction as you place the shoulder in extension and adduction. The humerus crosses over the midline to the right, the scapula is in anterior depression, and the palm faces toward the right ilium. A continuation of this motion would bring the patient into trunk flexion to the right.

Body Mechanics. Stand in a stride position by or above the patient's shoulder with your left foot forward. Face the line of motion. Start with the weight on your front foot and let the patient's motion push your weight to your back foot. Continue facing the line of motion.

Stretch. Apply the stretch to the shoulder and hand simultaneously. Your proximal hand does a rapid traction with rotation of the shoulder and scapula. Your distal hand gives traction to the wrist. *Caution:* Traction the wrist in line with the metacarpal bones. Do not force the wrist into more flexion.

Command. "Hand up, lift your arm." "Lift!"

Movement. The fingers and thumb extend as the wrist moves into radial extension. The radial side of the hand leads as the shoulder moves into flexion with abduction and external rotation. The scapula moves into posterior elevation. Continuation of this motion is an upward reach with elongation of the left side of the trunk.

Resistance. Your distal hand combines traction through the extended wrist with a rotary resistance for the radial deviation. Resistance to the forearm supination and shoulder external rotation and abduction comes from the rotary resistance at the wrist. The traction force resists the motions of wrist extension and shoulder flexion.

Your proximal hand combines a traction force with rotary resistance. The line of resistance is back toward the starting position. Maintaining the traction force will guide your resistance in the proper arc.

End Position. The humerus is in full flexion (about three fingers from the left ear), the palm facing about 45° to the coronal (lateral) plane. The scapula is fully elevated, adducted, and rotated upward. The elbow remains extended. The wrist is in full radial extension, fingers and thumb extended toward the radial side.

Timing for Emphasis. You may prevent motion in the beginning of the shoulder flexion and exercise the wrist, hand, or fingers.

7.1.3 Flexion-Abduction-External Rotation with Elbow Flexion (Fig. 7.3)

Joint	Movement	Muscles: principal components (Kendall and McCreary 1983)
Scapula	Posterior elevation	Trapezius, levator scapulae, serratus anterior
Shoulder	Flexion, abduction, external rotation	Deltoid (anterior), biceps (long head), coracobrachialis, supraspinatus, infraspinatus, teres minor
Elbow	Flexion	Biceps, brachialis
Forearm	Supination	Biceps, brachioradialis, supinator
Wrist	Radial extension	Extensor carpi radialis (longus and brevis)
Fingers	Extension, radial deviation	Extensor digitorum longus, interossei
Thumb	Extension, abduction	Extensor pollicis (longus and brevis), abductor pollicis longus

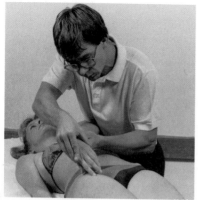

a

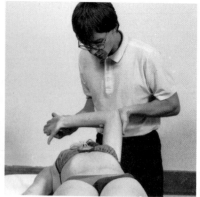

b

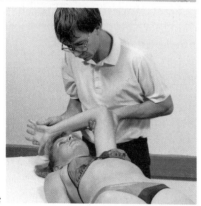

c

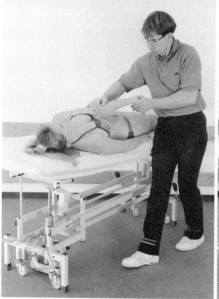

d

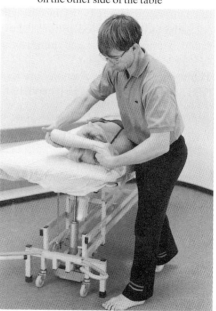

e

Fig. 7.3 a–e. Arm: flexion – abduction – external rotation with elbow flexion.
a–c Normal position of therapist.
d, e Alternative position with therapist on the other side of the table

72

Grip. *Distal hand:* Your distal grip is the same as described in Sect. 7.1.2 for the straight arm pattern.

Proximal hand: Your proximal hand may start with the grip used for the straight arm pattern. As the shoulder and elbow begin to flex, move this hand up to grip the humerus. You wrap your hand around the humerus from the medial side and use your fingers to give pressure opposite the direction of motion. The resistance to rotation comes from the line of your fingers and forearm.

Alternative Grip. The proximal hand may move to the scapula to emphasize that motion.

Elongated Position. Position the limb as for the straight arm pattern.

Body Mechanics. Stand in the same stride position as for the straight arm pattern. Allow the patient to push your weight from your front to back foot. Face the line of motion.

Alternative Position. You may stand on the right side of the table facing up toward the patient's left shoulder. Put your left hand on the patient's hand, your right hand on the humerus. Stand in a stride with your right leg forward. As the patient's arm moves up into flexion, step forward with your left leg. If you choose this position, move the patient to the right side of the table (Fig. 7.3 d, e).

Stretch. Use the same motions for the stretch reflex that you used with the straight arm pattern. *Caution:* Traction the wrist in line with the metacarpal bones. Do not force the wrist into more flexion

Command. "Hand up, lift your arm and bend your elbow." "Lift!"

Movement. The fingers and thumb extend and the wrist moves into radial extension as in the straight arm pattern. The elbow and shoulder motions begin next. The elbow flexion causes the hand and forearm to move across the face as the shoulder completes its flexion.

Resistance. Your distal hand resists the wrist and forearm as it did in the straight arm pattern. Add resistance to the elbow flexion by giving traction back toward the starting position. Your proximal hand combines rotary resistance with traction through the humerus toward the starting position. Give a separate force with each hand so that the resistance is appropriate for the strength of the shoulder and elbow.

End Position. The humerus is in full flexion with the scapula adducted, fully elevated, and rotated upward. The elbow is flexed, and the patient's forearm is touching his head. The wrist is again in full radial extension, fingers and thumb extended toward the radial side. The rotation in the shoulder and forearm are the same as in the straight arm pattern. If you extend the patient's elbow, the position is the same as in the straight arm pattern.

Timing for Emphasis. With three moving segments – shoulder, elbow and wrist – you may lock in any two and exercise the third.

With the elbow bent it is easy to exercise the external rotation separately from the other shoulder motions. Do these exercises where the strength of the shoulder flexion is greatest. You may work through the full range of shoulder external rotation during these exercises but return to the groove before finishing the pattern.

When exercising the patient's wrist or hand, move your proximal hand to the forearm and give resistance to the shoulder and elbow with that hand. Your distal hand is now free to give appropriate resistance to the wrist and hand motions. To exercise the fingers and thumb, move your proximal hand to give resistance just distal to the wrist. Your distal hand can now exercise the fingers, jointly or individually.

You may prevent motion in the beginning range of shoulder flexion and exercise the elbow, wrist, hand, or fingers.

7.1.4 Flexion-Abduction-External Rotation with Elbow Extension (Fig. 7.4)

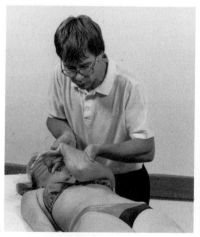
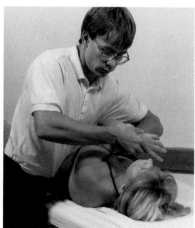

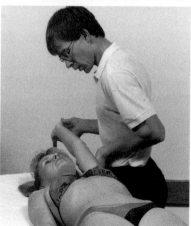

Fig. 7.4a–c. Arm: flexion – abduction – external rotation with elbow extension

74

Joint	Movement	Muscles: principal components (Kendall and McCreary 1983)
Scapula	Posterior elevation	Trapezius, levator scapulae, serratus anterior
Shoulder	Flexion, abduction, external rotation	Deltoid (anterior), biceps (long head), coracobrachialis, supraspinatus, infraspinatus, teres minor
Elbow	Extension	Triceps, anconeus
Forearm	Supination	Biceps, brachioradialis, supinator
Wrist	Radial extension	Extensor carpi radialis (longus and brevis)
Fingers	Extension, radial deviation	Extensor digitorum longus, interossei
Thumb	Extension, abduction	Extensor pollicis (longus and brevis), abductor pollicis longus

Grip. Your distal grip is the same as for the straight arm pattern. Wrap your proximal hand around the humerus from the medial side so that your fingers give pressure opposing the direction of motion.

Elongated Position. The positions of the patient's scapula, shoulder, forearm, and wrist are the same as for the straight arm pattern. The elbow is fully flexed.

Body Mechanics. Stand in the same stride position used for the straight arm pattern. Shift your weight as you did for the straight arm pattern.

Stretch. Apply the stretch to the shoulder, elbow, and hand simultaneously. Your proximal hand stretches the shoulder and scapula with a rapid traction and rotation motion. Your distal hand continues giving traction to the hand and wrist while stretching the elbow extension. *Caution:* Traction the wrist; do not force it into more flexion.

Command. "Hand up, push your arm up and straighten your elbow as you go." "Push!"

Movement. The fingers and thumb extend and the wrist moves into radial extension as before. The elbow and shoulder motions begin next. The elbow extension moves the hand and forearm in front of the face as the shoulder flexes. The elbow reaches full extension as the shoulder and scapula complete their motion.

Resistance. Your distal hand resists the wrist and forearm as in the straight arm pattern. Give resistance to the elbow extension by rotating the forearm and hand back towards the starting position of elbow flexion. Your proximal hand gives traction through the humerus combined with rotary resistance back toward the starting position.

Each of your hands uses the proper force to make the resistance appropriate for the strength of the elbow and the strength of the shoulder.

End Position. The end position is the same as the straight arm pattern.

Timing for Emphasis. Prevent elbow extension at the beginning of the range and exercise the shoulder. Lock in shoulder flexion in mid-range and exercise the elbow extension with supination.

7.1.5 Extension-Adduction-Internal Rotation (Fig. 7.5)

Joint	Movement	Muscles: principal components (Kendall and McCreary 1983)
Scapula	Anterior depression	Serratus anterior (lower), pectoralis minor, rhomboids
Shoulder	Extension, adduction, internal rotation	Pectoralis major, teres major, subscapularis
Elbow	Extended (position unchanged)	Triceps, anconeus
Forearm	Pronation	Brachioradialis, pronator (teres and quadratus)
Wrist	Ulnar flexion	Flexor carpi ulnaris
Fingers	Flexion, ulnar deviation	Flexor digitorum (superficialis and profundus), lumbricales, interossei
Thumb	Flexion, adduction, opposition	Flexor pollicis (longus and brevis), adductor pollicis, opponens pollicis

Grip. *Distal hand:* Your left hand contacts the palmar surface of the patient's hand. Your fingers are on the radial side (2nd metacarpal), your thumb gives counterpressure on the ulnar border (5th metacarpal). There is no contact on the dorsal surface. *Caution:* Do not squeeze the patient's hand.

Proximal hand: Your right hand comes from the radial side and holds the patient's forearm just proximal to the wrist. Your fingers contact the ulnar border, your thumb is on the radial border.

Elongated Position. Traction the entire arm and scapula while moving the wrist into radial extension. Continue the traction as you place the shoulder in flexion and abduction with the scapula in posterior elevation. The palm faces about 45° to the lateral plane. Continuation of the traction elongates the left side of the patient's trunk. Too much abduction prevents trunk elongation. Too much external rotation prevents the scapula from coming into full posterior elevation.

If the patient has just completed the antagonistic motion (flexion-abduction-external rotation), begin at the end of that pattern.

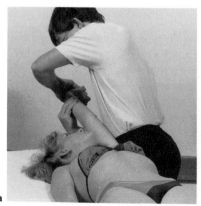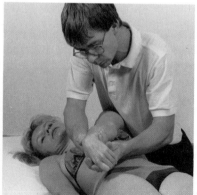

a   b

Fig. 7.5 a, b. Arm: extension – adduction – internal rotation

Body Mechanics. Stand in a stride position above the patient's shoulder with your left foot forward. Face the line of motion. Start with the weight on your back foot and let the patient's motion pull your weight to your front foot. Continue facing the line of motion.

Stretch. Apply the stretch to the shoulder and hand simultaneously. Your proximal hand does a rapid traction with rotation of the shoulder and scapula. Your distal hand gives traction to the wrist. *Caution:* Traction the wrist in line with the metacarpal bones. Do not force the wrist into more extension.

Command. "Squeeze my hand, pull down and across." "Squeeze and pull!"

Movement. The fingers and thumb flex as the wrist moves into ulnar flexion. The radial side of the hand leads as the shoulder moves into extension with adduction and internal rotation and the scapula into anterior depression. Continuation of this motion brings the patient into trunk flexion to the right.

Resistance. Your distal hand combines traction through the flexed wrist with rotary resistance to ulnar deviation. The rotary resistance at the wrist provides resistance to the forearm pronation and shoulder adduction and internal rotation. The traction at the wrist resists the wrist flexion and the shoulder extension.

Your proximal hand combines a traction force with rotary resistance. The line of resistance is back toward the starting position. Maintaining the traction force will guide your resistance in the proper arc.

Your hands may change from traction to approximation as the shoulder and scapula near the end of their range.

End Position. The scapula is in anterior depression. The shoulder is in extension, adduction, and internal rotation with the humerus crossing the midline to the right. The forearm is pronated, the wrist and fingers flexed with the palm facing toward the right ilium.

Timing for Emphasis. You may provent motion in the beginning range of shoulder extension and exercise the wrist, hand, and fingers. To exercise the fingers and thumb, move your proximal hand to give resistance just distal to the wrist. Your distal hand can now exercise the fingers, jointly or individually.

7.1.6 Extension-Adduction-Internal Rotation with Elbow Extension (Fig. 7.6)

Joint	Movement	Muscles: principal components (Kendall and McCreary 1983)
Scapula	Anterior depression	Serratus anterior (lower), pectoralis minor, rhomboids
Shoulder	Extension, adduction, internal rotation	Pectoralis major, teres major, subscapularis
Elbow	Extension	Triceps, anconeus
Forearm	Pronation	Brachioradialis, Pronator (teres and quadratus)
Wrist	Ulnar flexion	Flexor carpi ulnaris
Fingers	Flexion, ulnar deviation	Flexor digitorum (superficialis and profundus), lumbricales, interossei
Thumb	Flexion, adduction, opposition	Flexor pollicis (longus and brevis), adductor pollicis, opponens pollicis

Grip. Your distal grip is the same as for the straight arm pattern. Wrap your proximal hand around the humerus from underneath so that your fingers can give pressure opposite the direction of rotation.

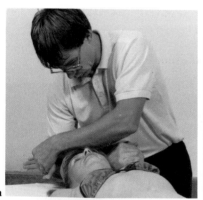
a

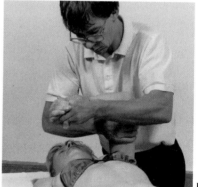
b

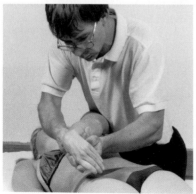
c

Fig. 7.6 a–c. Arm: extension – adduction – internal rotation with elbow extension

Elongated Position. The humerus is in full flexion with the scapula adducted, elevated, and rotated upward. The elbow is flexed, the patient's forearm is touching his or her head. The wrist is in full radial extension with the fingers extended.

Body Mechanics. Your body mechanics are the same as for the straight arm pattern.

Stretch. Apply the stretch to the shoulder and hand simultaneously. Your proximal hand does a rapid traction with rotation of the shoulder and scapula. Your distal hand gives traction to the wrist. In most cases the patient's forearm is touching his or her head, preventing increased elbow flexion.

Command. "Squeeze my hand, push down and across, straighten your elbow." "Squeeze and push!"

Movement. The fingers and thumb flex and the wrist moves into ulnar flexion. The shoulder begins its motion into extension-adduction, and then the elbow begins to extend, the hand moving down toward the opposite hip. The elbow reaches full extension as the shoulder and scapula complete their motion.

Resistance. Your proximal hand gives traction through the humerus combined with rotary resistance back toward the starting position. Your distal hand resists the wrist and forearm, as in the straight arm pattern. Give resistance to the elbow extension by rotating the forearm and hand back towards the starting position of elbow flexion.

Each of your hands uses the proper force to make the resistance appropriate for the strength of the elbow and the strength of the shoulder.

End Position. The end position is the same as the straight arm pattern.

Timing for Emphasis. Prevent elbow extension at the beginning of the range and exercise the shoulder. Lock in shoulder extension in mid-range and exercise the elbow extension with pronation.

7.1.7 Extension-Adduction-Internal Rotation with Elbow Flexion (Fig. 7.7)

Joint	Movement	Muscles: principal components (Kendall and McCreary 1983)
Scapula	Anterior depression	Serratus anterior (lower), pectoralis minor, rhomboids,
Shoulder	Extension, adduction, internal rotation	Pectoralis major, teres major, subscapularis
Elbow	Flexion	Biceps, brachialis
Forearm	Pronation	Brachioradialis, pronator (teres and quadratus)
Wrist	Ulnar flexion	Flexor carpi ulnaris
Fingers	Flexion, ulnar deviation	Flexor digitorum (superficialis and profundus), lumbricales, interossei
Thumb	Flexion, adduction, opposition	Flexor pollicis (longus and brevis), adductor pollicis, opponens pollicis

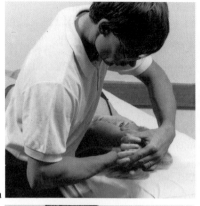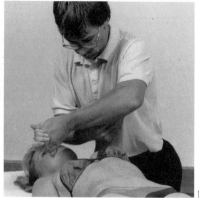

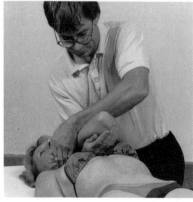

Fig. 7.7 a–c. Arm: extension – adduction –
internal rotation with elbow flexion

Grip. Your distal and proximal grips are the same as the ones used for extension-adduction-internal rotation with elbow extension.

Alternative Grip. After the motion begins, the proximal hand may move to the scapula to emphasize that motion.

Elongated Position. The position is the same as the straight arm pattern.

Stretch. The stretch is the same as for the straight arm pattern.

Command. "Squeeze my hand, pull down and across and bend your elbow." "Squeeze and pull."

Movement. The fingers and thumb flex and the wrist moves into ulnar flexion. The shoulder starts into extension-adduction and the elbow begins to flex. The elbow reaches full flexion as the shoulder and scapula complete their motion.

End Position. The humerus is in extension with adduction, the scapula in anterior depression. The elbow is fully flexed. The wrist is in ulnar flexion and the hand closed. The amount of rotation in the shoulder and forearm is the same as in the straight arm pattern. If you extend the patient's elbow, the position is the same as the straight arm pattern.

Timing for Emphasis. With the three moving segments – shoulder, elbow and wrist – you may again lock in any two and exercise the third.

With the elbow bent it is easy to exercise the internal rotation separately from the other motions. Exercise this motion where the strength of shoulder extension is greatest. You may work through the full range of shoulder internal rotation during these exercises and return to the groove before finishing the pattern.

When exercising the patient's wrist or hand, move your proximal hand to the forearm and give resistance to the shoulder and elbow by pulling back toward the starting position. Your distal hand is now free to give appropriate resistance to the wrist and hand. To exercise the fingers and thumb, move your proximal stabilizing hand just distal to the wrist.

You may prevent motion in the beginning range of shoulder extension and exercise the elbow, wrist, hand, or fingers. In addition, you may lock in the shoulder and elbow motion in mid-range to exercise the wrist and hand. This places the hand where the patient can see it as it moves.

7.1.8 Flexion-Adduction-External Rotation (Fig. 7.8)

Joint	Movement	Muscles: principal components (Kendall and McCreary 1983)
Scapula	Anterior elevation	Serratus anterior (upper), trapezius
Shoulder	Flexion, adduction, external rotation	Pectoralis major (upper) deltoid (anterior), biceps, coracobrachialis
Elbow	Extended (position unchanged)	Triceps, anconeus
Forearm	Supination	Brachioradialis, supinator
Wrist	Radial flexion	Flexor carpi radialis
Fingers	Flexion, radial deviation	Flexor digitorum (superficialis and profundus), lumbricales, interossei
Thumb	Flexion, adduction	Flexor pollicis (longus and brevis), adductor pollicis

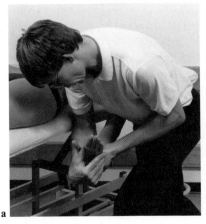 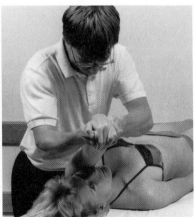

a   b

Fig. 7.8 a, b. Arm: flexion – adduction – external rotation

Grip. *Distal hand:* Your right hand contacts the palmar surface of the patient's hand. Your fingers are on the ulnar side (5th metacarpal), your thumb gives counterpressure on the radial side (2nd metacarpal). There is no contact on the dorsal surface. *Caution:* Do not squeeze the hand.

Proximal hand: Your left hand grips the patient's forearm from underneath just proximal to the wrist. Your fingers are on the radial side, your thumb on the ulnar side.

Alternative Grip. To emphasize shoulder or scapula motions, move the proximal hand to grip the upper arm or the scapula after the shoulder begins its motion.

Elongated Position. Traction the entire arm and scapula while moving the wrist into ulnar extension. Continue the traction as you place the shoulder in extension and adduction with the scapula in posterior depression. The palm faces about 45° in toward the body. Continuation of the traction shortens the left side of the patient's trunk. Too much shoulder abduction prevents the trunk motion and pulls the scapula out of position. Too much internal rotation tilts the scapula forward.

Body Mechanics. Stand in a stride position by the patient's elbow, facing towards the patient's feet. The patient's motion of flexion with external rotation pivots you around so you face diagonally up toward the patient's head. Let the patient's motion pull your weight from your back to your front foot.

Stretch. Your proximal hand does a rapid traction with rotation of the shoulder and scapula. At the same time your distal hand gives traction to the wrist. *Caution:* Traction the wrist in line with the metacarpal bones. Do not force the wrist into more extension.

Command. "Squeeze my hand, pull up and across your nose." "Squeeze and pull."

Movement. The fingers and thumb flex as the wrist moves into radial flexion. The radial side of the hand leads as the shoulder moves into flexion with adduction and external rotation and the scapula into anterior elevation. Continuation of this motion elongates the patient's trunk with rotation toward the right.

Resistance. Your distal hand combines traction through the flexed wrist with rotary resistance to radial deviation. The rotary resistance at the wrist provides resistance to the forearm supination and to the shoulder adduction and external rotation. The traction force resists both the wrist flexion and shoulder flexion.

Your proximal hand combines a traction force with rotary resistance. The line of resistance is back toward the starting position. Maintaining the traction force guides your resistance in the proper arc.

End Position. The scapula is in anterior elevation, and the shoulder is in flexion and adduction with external rotation so the humerus crosses at the midline over the patient's face. The forearm is supinated, the elbow straight, and the wrist and fingers flexed.

Timing for Emphasis. You may prevent motion in the beginning range of shoulder flexion and exercise the wrist, hand, or fingers. Lock in the forearm rotation or allow it to move with the wrist.

7.1.9 Flexion-Adduction-External Rotation with Elbow Flexion (Fig. 7.9)

Joint	Movement	Muscles: principal components (Kendall and McCreary 1983)
Scapula	Anterior elevation	Serratus anterior (upper), trapezius
Shoulder	Flexion, adduction, external rotation	Pectoralis major (upper), deltoid (anterior), biceps (long head), coracobrachialis
Elbow	Flexion	Biceps, brachialis
Forearm	Supination	Brachioradialis, biceps, Supinator
Wrist	Radial flexion	Flexor carpi radialis
Fingers	Flexion, radial deviation	Flexor digitorum (superficialis and profundus), lumbricales, interossei
Thumb	Flexion, adduction	Flexor pollicis (longus and brevis), adductor pollicis

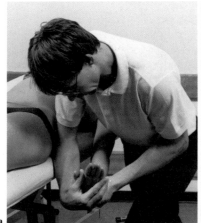

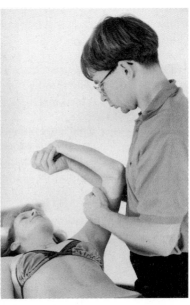

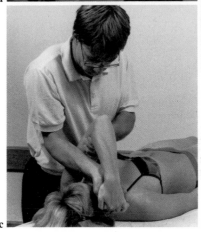

Fig. 7.9 a–c. Arm: flexion – adduction – external rotation with elbow flexion

Grip. Your distal grip is the same as used for the straight arm pattern. Your proximal hand may start with the grip for the straight arm pattern. As the shoulder and elbow begin to flex, move this hand up to grip the humerus. Wrap your hand around the humerus from the medial side and use your fingers to give pressure opposite the direction of motion. The resistance to rotation comes from the line of your fingers and forearm (Fig. 7.9 b).

Alternative Grip. The proximal hand may move to the scapula to emphasize that motion.

Elongated Position. Position the limb as for the straight arm pattern.

Body Mechanics. Your body mechanics are the same as for the straight arm pattern. Use your body weight for resistance.

Stretch. Use the same motions for the stretch reflex that you used with the straight arm pattern. *Caution:* Traction the wrist in line with the metacarpal bones. Do not force the wrist into more flexion

Command. "Squeeze my hand, pull up across your nose and bend your elbow." "Squeeze and pull."

Movement. After the wrist flexes and the forearm supinates, the shoulder and elbow begin to flex. The shoulder and elbow move at the same speed and complete their movements at the same time.

Resistance. Your distal hand resists the wrist and forearm as in the straight arm pattern. Added is the rotary resistance to elbow flexion. Your proximal hand rotates and tractions the humerus back toward the starting position. You give a separate force with each hand so that the resistance is appropriate for the strength of the shoulder and elbow.

End Position. The patient's shoulder, forearm, and hand are positioned as in the straight arm pattern. The elbow is flexed, and the patient's fist may touch his or her right ear. The rotations in the shoulder and forearm are the same as in the straight arm pattern. Extend the elbow to check the amount of rotation.

Timing for Emphasis. With three moving segments – shoulder, elbow and wrist – you may lock in any two and exercise the third.

As with the other bent elbow pattern combinations it is easy to exercise the external rotation separately from the other shoulder motions. Do this where the strength of the shoulder flexion is greatest. If you work through the full range of shoulder external rotation, return to the groove before finishing the pattern.

When exercising the wrist or hand, move your proximal hand to the forearm or hand to stabilize and resist the proximal joints. Your other hand grips distal to the joints being exercised.

You may prevent motion in the beginning range of shoulder flexion and exercise the elbow flexion with supination, the wrist, or the fingers.

7.1.10 Flexion-Adduction-External Rotation with Elbow Extension (Fig. 7.10)

Joint	Movement	Muscles: principal components (Kendall and McCreary 1983)
Scapula	Anterior elevation	Serratus anterior (upper), trapezius
Shoulder	Flexion, adduction, external rotation	Pectoralis major (upper), deltoid (anterior), biceps, coracobrachialis
Elbow	Extension	Triceps, anconeus
Forearm	Supination	Brachioradialis, supinator
Wrist	Radial flexion	Flexor carpi radialis
Fingers	Flexion, radial deviation	Flexor digitorum (superficialis and profundus), lumbricales, interossei
Thumb	Flexion, adduction	Flexor pollicis (longus and brevis), adductor pollicis

Grip. *Distal hand:* The distal grip is the same as used for the straight arm pattern.

Proximal hand: Your proximal hand starts with the grip on the forearm used with the straight arm pattern. As the shoulder begins to flex and the elbow to extend, this hand can move up to grip the humerus. In this case wrap your hand around the

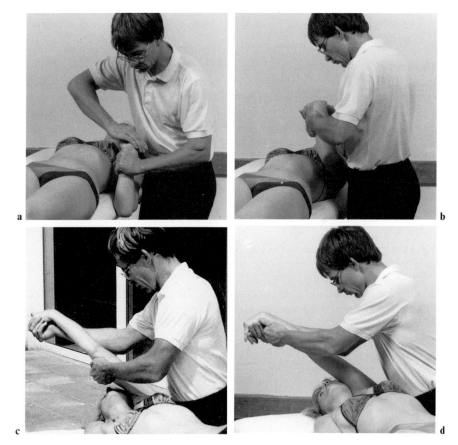

Fig. 7.10 a–d. Arm: flexion – adduction – external rotation with elbow extension

humerus from the medial side and use your fingers to give pressure opposite the direction of motion. You may use the grip on the humerus from the start of the pattern.

Elongated Position. Start by positioning the limb as you did for the straight arm pattern. Maintaining traction on the shoulder and scapula with your proximal hand, use that hand to flex the elbow. Your distal hand tractions the wrist into ulnar extension. If you begin with your left hand on the humerus, your distal (right) hand flexes the elbow.

Body Mechanics. Your body mechanics are the same as for the straight arm pattern. Use your body weight for resistance.

Stretch. Your proximal hand does a rapid traction with rotation of the shoulder and scapula. At the same time your distal hand gives traction to the wrist.
Caution: Do not force the wrist into more extension.

Command. "Squeeze my hand, push up and across your nose and straighten your elbow." "Squeeze and push!"

Movement. After the wrist flexes and the forearm supinates, the shoulder begins to flex and elbow to extend. The shoulder and elbow should complete their motions at the same time.

Resistance. Your distal hand resists the wrist and forearm as in the straight arm pattern. Added is the rotary resistance to elbow extension. Your proximal hand rotates and tractions the humerus back toward the starting position.

You give a separate force with each hand so that the resistance is appropriate for the strength of the shoulder and elbow.

End Position. The patient's shoulder, forearm, and hand are positioned as in the straight arm pattern.

Timing for Emphasis. The emphasis here is to teach the patient to combine shoulder flexion with elbow extension in a smooth motion.

7.1.11 Extension-Abduction-Internal Rotation (Fig. 7.11)

Joint	Movement	Muscles: principal components (Kendall and McCreary 1986)
Scapula	Posterior depression	Rhomboids
Shoulder	Extension, abduction, internal rotation	Latissimus dorsi, deltoid (middle, posterior), triceps, teres major, subscapularis
Elbow	Extended (position unchanged)	Triceps, anconeus
Forearm	Pronation	Brachioradialis, pronator (teres and quadratus)
Wrist	Ulnar extension	Extensor carpi ulnaris
Fingers	Extension, ulnar deviation	Extensor digitorum longus, lumbricales, interossei
Thumb	Palmar abduction, extension	Abductor pollicis (brevis)

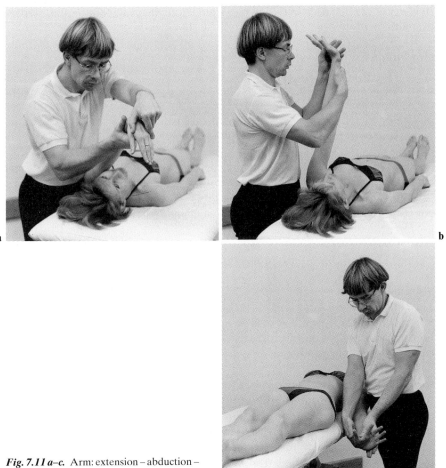

Fig. 7.11 a–c. Arm: extension – abduction – internal rotation

Grip. *Distal hand:* Your left hand grips the dorsal surface of the patient's hand. Your fingers are on the ulnar side (5th metacarpal), your thumb gives counter-pressure on the radial side (2nd metacarpal). There is no contact on the palm. *Caution:* Do not squeeze the hand.

Proximal hand: With your hand facing the ventral surface, use the lumbrical grip to hold the radial and ulnar sides of the patient's forearm proximal to the wrist.

Alternative Grip. To emphasize shoulder or scapula motions, move the proximal hand to the upper arm or to the scapula after the shoulder begins to extend.

Elongated Position. Traction the entire arm and scapula while you move the wrist into radial flexion and the forearm into supination. Continue the traction as you place the shoulder in flexion and adduction and the scapula in anterior elevation. The humerus crosses over the patient's nose and the palm faces toward the patient's right ear. A continuation of this motion would bring the patient into trunk elongation with rotation to the right.

If the patient has just completed the antagonistic motion (flexion-adduction-external rotation), begin at the end of that pattern.

Body Mechanics. Stand in a stride position in the line of the motion facing toward the patient's hand. Start with the weight on your front foot and let the patient's motion push your weight to your back foot. As the patient's arm nears the end of the range, your body turns so you face the patient's feet.

Stretch. Apply the stretch to the shoulder and hand simultaneously. Your proximal hand does a rapid traction with rotation of the shoulder and scapula. Your distal hand gives traction to the wrist. Combine this motion with traction to the wrist with your distal hand. *Caution:* Traction the wrist in line with the metacarpal bones. Do not force the wrist into more flexion.

Command. "Hand back, push your arm down to your side." "Push!"

Movement. The fingers and thumb extend as the wrist moves into ulnar extension. The ulnar side of the hand leads as the shoulder moves into extension with abduction and internal rotation. The scapula moves into posterior depression. Continuation of this motion is a downward reach toward the back of the heel with shortening of the left side of the trunk.

Resistance. Your distal hand combines traction through the extended wrist with a rotary resistance for the ulnar deviation. The resistance to the forearm pronation and shoulder internal rotation and abduction comes from the rotary resistance at the wrist. The traction force resists the motions of wrist and shoulder extension.

Your proximal hand combines a traction force with rotary resistance. The line of resistance is back toward the starting position.

As the arm nears the end of the range of extension, change from traction to approximation.

End Position. The scapula is in full posterior depression. The humerus is in extension by the left side, the forearm is pronated, and the palm is facing about 45° to the lateral plane. The wrist is in ulnar extension, the fingers are extended toward the ulnar side, and the thumb is extended and abducted at right angles to the palm.

Timing for Emphasis. You may prevent motion in the beginning of the shoulder extension and exercise the wrist, hand, or fingers. This positions the hand where the patient can see it during the exercise.

7.1.12 Extension-Abduction-Internal Rotation with Elbow Extension (Fig. 7.12)

Joint	Movement	Muscles: principal components (Kendall and McCreary 1983)
Scapula	Posterior depression	Rhomboids
Shoulder	Extension abduction, internal rotation	Latissimus dorsi, deltoid (middle, posterior), triceps, teres major, subscapularis
Elbow	Extension	Triceps, anconeus
Forearm	Pronation	Brachioradialis, pronator (teres and quadratus)
Wrist	Ulnar extension	Extensor carpi ulnaris
Fingers	Extension, ulnar deviation	Extensor digitorum longus, Lumbricales, interossei
Thumb	Palmar abduction, extension	Abductor pollicis (brevis)

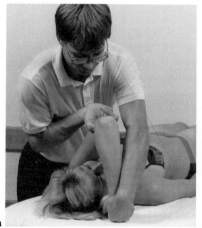
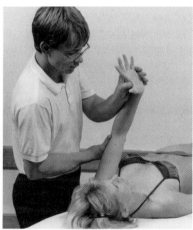
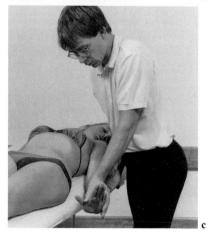

Fig. 7.12 a–c. Arm: extension – abduction – internal rotation with elbow extension

Grip. *Distal hand:* Your distal grip is the same as used for the straight arm pattern.

Proximal hand: Wrap your hand around the humerus so your fingers can give pressure opposite the direction of internal rotation.

Elongated Position. The positions of the scapula, shoulder, forearm, and wrist are the same as for the straight arm pattern. The patient's elbow is fully flexed.

Body Mechanics. Your body mechanics are the same as for the straight arm pattern.

Stretch. Apple the stretch to the shoulder, elbow, and hand simultaneously. The stretch of the shoulder comes from a rapid traction with rotation of the shoulder and scapula by the proximal hand. The distal hand continues giving traction to the hand and wrist while rapidly increasing the elbow flexion. *Caution:* Traction the wrist; do not force it into more flexion.

Command. "Hand up, push your arm down toward me and straighten your elbow as you go." "Push!"

Movement. The fingers extend and the wrist moves into ulnar extension. The shoulder begins its motion into extension-abduction, and then the elbow begins to extend. The elbow reaches full extension as the shoulder and scapula complete their motion.

Resistance. Your proximal hand gives traction through the humerus combined with rotary resistance back toward the starting position. Your distal hand resists the wrist and forearm as in the straight arm pattern. Give resistance to the elbow extension by rotating the forearm and hand back towards the starting position of elbow flexion. When the shoulder and elbow near full extension, change from traction to approximation.

End Position. The end position is the same as the straight arm pattern.

Timing for Emphasis. Prevent elbow extension at the beginning of the range and exercise the shoulder. Prevent shoulder extension at the beginning of the range and exercise the elbow extension with pronation. Lock in shoulder extension in mid-range and exercise the elbow extension with pronation.

7.1.13 Extension-Abduction-Internal Rotation with Elbow Flexion
(Fig. 7.13)

Joint		Muscles: principal components (Kendall and McCreary 1983)
Scapula	Posterior depression	Rhomboids
Shoulder	Extension, abduction, internal rotation	Latissimus dorsi, deltoid (middle, posterior), triceps, teres major, subscapularis
Elbow	Flexion	Biceps, brachialis
Forearm	Pronation	Brachioradialis, pronator (teres and quadratus)
Wrist	Ulnar extension	Extensor carpi ulnaris
Fingers	Extension, Ulnar deviation	Extensor digitorum longus, lumbricales, interossei
Thumb	Palmar abduction, extension	Abductor pollicis (brevis)

Grip. *Distal hand:* Your distal grip is the same as used for the straight arm pattern.

Proximal hand: Your proximal hand may start with the grip on the forearm. As the shoulder and elbow motions begin, wrap your proximal hand around the humerus from underneath. Your fingers give pressure opposite the direction of rotation and resist the shoulder extension.

Alternative Grip. The proximal hand may move to the scapula to emphasize that motion.

Elongated Position. The position is the same as for the straight arm pattern.

Body Mechanics. You may stand on the opposite side of the table. Face the diagonal and use your body weight for resistance (Fig. 7.14).

Stretch. The stretch is the same as for the straight arm pattern.

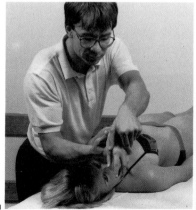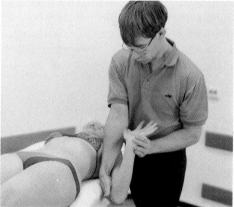

Fig. 7.13 a, b. Arm: extension – abduction – internal rotation with elbow flexion

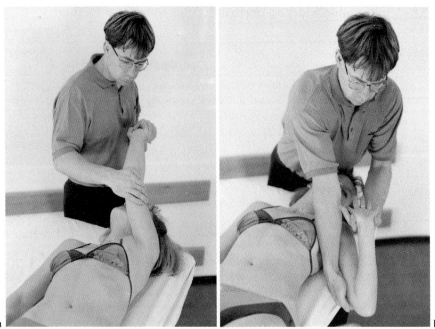

a   b

Fig. 7.14 a, b. Arm: extension – abduction – internal rotation with elbow flexion: therapist at the head end of the table

Command. "Fingers and wrist back, push down and out and bend your elbow." "Push down and bend your elbow."

Movement. The fingers extend and the wrist moves into ulnar extension. The shoulder begins its motion into extension-abduction, then the elbow begins to flex. The elbow reaches full flexion as the shoulder and scapula complete their motion.

Resistance. The distal hand gives the same resistance to the shoulder movement as in the straight arm pattern and a flexion resistance for the elbow. The proximal hand, at the beginning on the forearm, gives the same resistance as with the straight arm pattern. As soon as the proximal hand is on the upper arm, it gives resistance to rotation and shoulder extension. You can change the traction into approximation at the end of the movement.

End Position. The scapula is in posterior depression, the humerus in extension with abduction. The elbow is fully flexed. The wrist is again in ulnar extension and the hand open. The rotation in the shoulder and forearm are the same as in the straight arm pattern.

Timing for Emphasis. Lock in the wrist extension and elbow flexion, then exercise the shoulder in hyperextension and the scapula in posterior depression. When elbow flexion is stronger than extension, use this combination to exercise the patient's wrist and fingers.

7.2 Thrust and Withdrawal Patterns

In the upper extremity patterns certain combinations of motions are fixed. The shoulder and forearm rotate in the same direction, supination occurs with external rotation and pronation with internal rotation. Extension of the hand and wrist is combined with shoulder abduction, flexion of the hand and wrist with shoulder adduction. In the *thrust combinations* the shoulder and forearm rotate in opposite directions. In thrusting the extension motions of the hand and wrist are associated with shoulder adduction. Thrust reversals (withdrawal) combine flexion of the fingers and wrist with shoulder abduction. The elbow extends in the thrust patterns and flexes in the thrust reversals. The muscles used in these activities are the same as those used in the patterns.

Therapist Position. The therapist's position remains in the line of the motion. Because of the "pushing" and "pulling" motions of the thrust-withdrawal diagonals, an effective position is at the opposite side of the patient. This position is illustrated with both thrust diagonals.

Grips. The distal and proximal grips are those used to resist the same distal and proximal pattern movements.

Timing. The sequencing of the movements is the same as it is in the patterns. The hand and wrist complete their motion, and then the elbow, shoulder and scapula move through their ranges together.

Timing for Emphasis. The thrust and withdrawal patterns are exercised as a unit. Do the patterns singly or in combinations. Lock in the strong arm to reinforce the work of the weaker arm. Combination of isotonics and dynamic reversals (slow reversals) work well with these combinations.

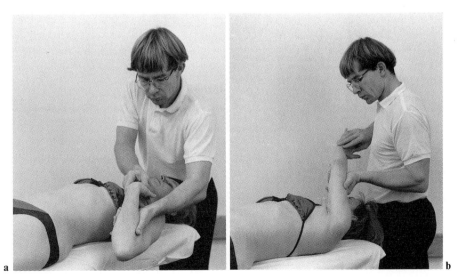

a   b

Fig. 7.15 a, b. Arm: ulnar thrust

7.2.1 Ulnar Thrust and Withdrawal

Ulnar Thrust (Fig. 7.15). The wrist and fingers extend with ulnar deviation. The shoulder moves in the pattern of flexion-adduction-external rotation with scapular anterior elevation and the elbow extends with forearm pronation.

Withdrawal from Ulnar Thrust (Fig. 7.16). The wrist and fingers flex with radial deviation. The shoulder moves in the pattern of extension-abduction-internal rotation with scapular posterior depression and the elbow flexes with forearm supination.

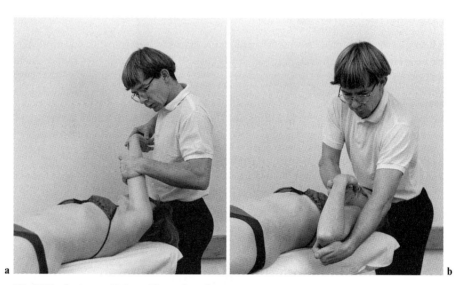

a   b

Fig. 7.16 a, b. Arm: withdrawal from ulnar thrust

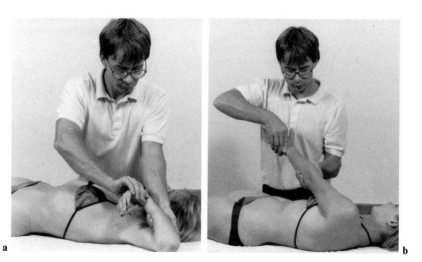

a   b

Fig. 7.17 a, b. Arm: radial thrust

7.2.2 Radial Thrust and Withdrawal

Radial Thrust (Fig. 7.17). The wrist and fingers extend with radial deviation. The shoulder moves in the pattern of extension-adduction-internal rotation with scapular anterior depression and the elbow extends with forearm supination.

Withdrawal from Radial Thrust (Fig. 7.18). The wrist and fingers flex with ulnar deviation. The shoulder moves in the pattern of flexion-abduction-external rotation with scapular posterior elevation and the elbow flexes with forearm pronation.

7.3 Bilateral Arm Patterns

Bilateral arm work allows you to use irradiation from the patient's strong arm to facilitate weak motions or muscles in the involved arm. You can use any combination

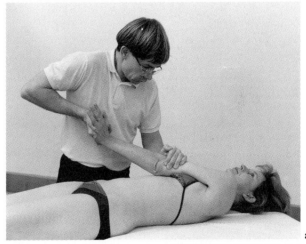

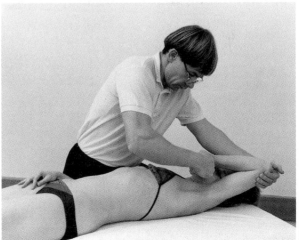

Fig. 7.18 a, b. Arm: withdrawal from radial thrust

of patterns in any position. Work with those that give you and the patient the greatest advantage in strength and control.

When you exercise both arms at the same time there is always more demand on the trunk muscles than when only one arm is exercising. You can increase this demand on the trunk by putting your patient in less supported positions such as sitting, kneeling, or standing. Here we show all the bilateral arm patterns for a supine patient to show more clearly the therapist's body position and grips.

Bilateral Symmetrical. Flexion-abduction-external rotation (Fig. 7.19).

Bilateral Asymmetrical. Flexion-abduction-external rotation with the right arm, flexion-adduction-external rotation with the left arm (Fig. 7.20).

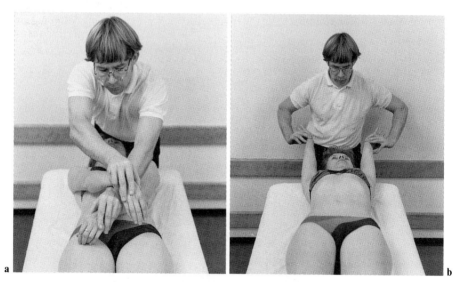

a   b

Fig. 7.19 a, b. Arm: bilateral symmetrical pattern of flexion – abduction – external rotation

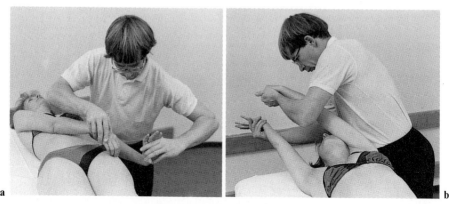

a   b

Fig. 7.20 a, b. Arm: bilateral asymmetrical pattern of flexion – abduction on the right and flexion – adduction on the left

96

Bilateral Symmetrical Reciprocal. Flexion-abduction-external rotation with the right arm, extension-adduction-internal rotation with the left arm (Fig. 7.21).

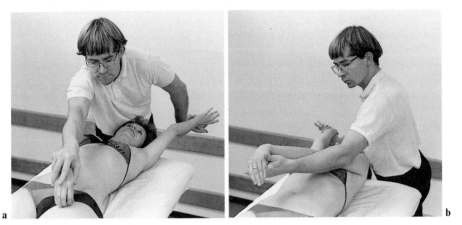

Fig. 7.21 a, b. Arm: bilateral symmetrical reciprocal pattern with flexion – abduction – external rotation on the right and extension – adduction – internal rotation on the left

Fig. 7.22 a–c. Arm: bilateral asymmetrical reciprocal pattern with extension – adduction on the right and flexion – adduction on the left

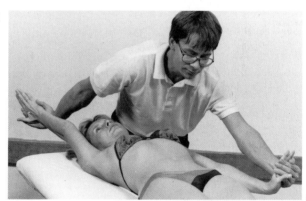

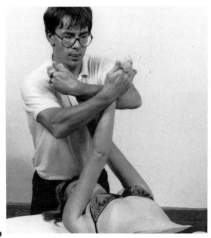

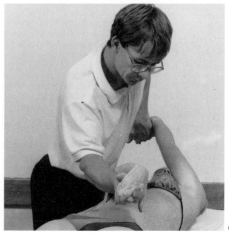

Bilateral Asymmetrical Reciprocal. Extension-adduction-internal rotation with the right arm, flexion-adduction-external rotation with the left arm (Fig. 7.22).

7.4 Changing the Patient's Position

There are many advantages to exercising the patient's arms in a variety of positions. These include letting the patient see the arm, adding or eliminating the effect of gravity from a motion, and working with functional motions in functional positions. There are also disadvantages for each position. Choose the positions that give the desired benefits with the fewest drawbacks.

7.4.1 Arm Patterns in a Sidelying Position

In this position the patient is free to move and stabilize the scapula without interference from the supporting surface. You may stabilize the patient's trunk with external support or the patient may do the work of stabilizing the trunk.

Extension-abduction-internal rotation is shown in Fig. 7.23 a (elongated position) and Fig. 7.23 b (end range position).

7.4.2 Arm Patterns Lying Prone on Elbows

Working with the patient in this position allows you to exercise the end range of the shoulder abduction patterns against gravity. The patient must bear weight on the other shoulder and scapula and maintain the head against gravity while exercising.

Flexion-abduction-external rotation at end range is shown in Fig. 7.24.

 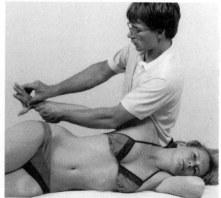

a   b

Fig. 7.23 a, b. Arm pattern in a sidelying position: extension – abduction – internal rotation. *a* Elongated position, *b* end position

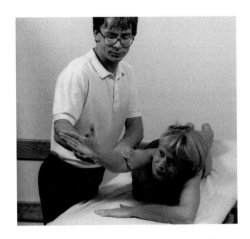

Fig. 7.24. Arm pattern lying prone on the elbows: flexion – abduction – external rotation at the end position

7.4.3 Arm Patterns in a Sitting Position

In this position you can exercise the patient's arms through their full range or limit the work to functional motions such as eating, reaching, dressing. Bilateral arm patterns may be done to challenge the patient's balance and stability (Fig. 7.25).

Fig. 7.25. Arm pattern in a sitting position: flexion – abduction – external rotation

7.4.4 Arm Patterns in a Crawling Position

Working in this position the patient must stabilize the trunk and bear weight on one arm while moving the other. As in the prone position, the shoulder flexor muscles work against gravity (Fig. 7.26). *Caution:* Do not allow the spine to move into undesired positions or postures.

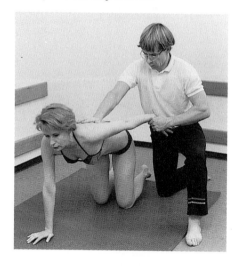

Fig. 7.26. Arm pattern in a crawling position: extension – abduction – internal rotation

Fig. 7.27. Arm pattern in a kneeling position: irradiation from the arm pattern flexion – abduction – external rotation for extension of the trunk and hip

7.4.5 Arm Patterns in a Kneeling Position

Working in this position requires the patient to stabilize both the trunk, hips and knees while doing arm exercises (Fig. 7.27). *Caution:* Do not allow the spine to move into undesired positions or postures.

Reference

Kendall FP, McCreary EK (1983) Muscles, Testing and Function. Williams and Wilkins, Baltimore

8 The Lower Extremity

8.1 Leg Patterns

8.1.1 Introduction

Lower extremity patterns are used to treat dysfunctions in the leg and foot caused by muscular weakness, incoordination, and joint restrictions. The leg patterns are also used to exercise the trunk. Resistance to strong leg muscles produces irradiation to weaker muscles elsewhere in the body.

8.1.1.1 Diagonal Motion

The leg has two diagonals:

1. Flexion-abduction-internal rotation and extension-adduction-external rotation
2. Flexion-adduction-external rotation and extension-abduction-internal rotation

The leg moves through the diagonals in a straight line with the rotation occurring smoothly throughout the motion. In the normal timing of the pattern, the foot and ankle move through their full range first, and the other joints then move through their ranges together.

The basic patterns of the left leg with the subject supine are shown in Fig. 8.1. All descriptions refer to this arrangement. To work with the right leg just change the word "left" to "right" in the instructions. Variations of position are shown later in the chapter.

8.1.1.2 Patient Position

Position the patient close to the edge of the table. The patient's spine should be in a neutral position without side-bending or rotation. Before beginning a lower extremity pattern, hold the patient's leg in a middle position where the lines of the two diagonals cross. The hip should be in neutral rotation. From this midline position, move the extremity into the elongated range of the pattern.

8.1.1.3 Therapist Position

The therapist stands on the left side of the table facing the line of the diagonal, arms and hands aligned with the motion. All grips described in the first part of this chapter (Sect. 8.1) assume that you are in this position.

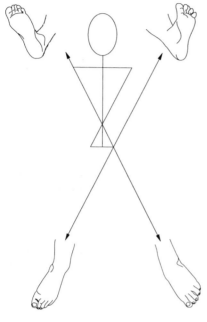

Hip: Flexion adduction
external rotation

Ankle: dorsi flexion
Foot: inversion
Toes: Extension (medial)

Hip: Flexion abduction
internal rotation

Ankle: dorsi flexion
Foot: eversion
Toes: Extension (lateral)

Hip: Extension adduction
external rotation

Ankle: plantar flexion
Foot: inversion
Toes: flexion (medial)

Hip: Extension abduction
internal rotation

Ankle: plantar flexion
Foot: eversion
Toes: flexion (lateral)

Fig. 8.1. Lower extremity
diagonals. (Courtesy of V. Jung)

We first give the basic position and body mechanics for exercising the straight leg pattern. When we describe variations in the patterns we identify any changes in position or body mechanics. Some of these variations are pictured at the end of the chapter.

8.1.1.4 Grips

The grips follow the basic procedures for manual contact, that is, opposite the direction of movement. The first part of this chapter (Sect. 8.1) describes the two-handed grip used when the therapist stands next to the moving lower extremity. The basic grip is described for each straight leg pattern. The grips are modified when the therapist's or patient's position is changed. The grips also change when the therapist uses only one hand while the other hand controls another pattern or extremity.

The grip on the foot contacts the active surface, dorsal or plantar, and holds the sides of the foot to resist the rotary components. Using the lumbrical grip will prevent squeezing or pinching the patient's foot. Remember, pain inhibits effective motion.

8.1.1.5 Resistance

The direction of the resistance is an arc back toward the starting position. The angle of the therapist's hands and arms giving the resistance changes as the limb moves through the pattern. Traction or approximation is an important part of the resistance.

8.1.1.6 Normal Timing

The foot and ankle (distal component) begin the pattern by moving through their full range. Rotation at the hip and knee accompanies the rotation (eversion or inversion) of the foot. After the distal movement is completed, the hip or hip and knee move jointly through their range.

In the sections on *timing for emphasis* we offer some suggestions for exercising components of the patterns. Any of the techniques may be used. We have found that repeated stretch (repeated contractions) and combination of isotonics work well, but do not limit yourself to the exercises we suggest: use your imagination.

8.1.2 Flexion-Abduction-Internal Rotation (Fig. 8.2)

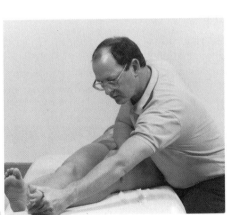
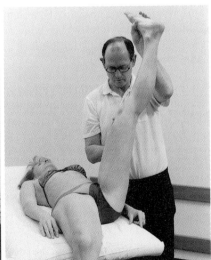

a   b

Fig. 8.2 a, b. Leg: flexion – abduction – internal rotation

Joint	Movement	Muscles: principal components (Kendall and McCreary 1983)
Hip	Flexion, abduction, internal rotation	Tensor fascia lata, rectus femoris, gluteus medius (anterior), gluteus minimus
Knee	extended (position unchanged)	Quadriceps
Ankle/foot	Dorsiflexion, eversion	Peroneus tertius
Toes	Extension, deviation to the left	extensor hallucis, extensor digitorum

Grip. *Distal hand:* Your left hand grips the dorsum of the patient's foot. Your fingers are on the lateral border and your thumb gives counterpressure on the medial border. Hold the sides of the foot but don't put any contact on the plantar surface. To avoid blocking toe motion, keep your grip proximal to the metatarsal-phalangeal joints. *Caution:* Do not squeeze or pinch the foot.

Proximal hand: Place your right hand on the anterior-lateral surface of the thigh just proximal to the knee. The fingers are on the top, the thumb on the lateral surface.

Elongated Position. Traction the entire limb while you move the foot into plantar flexion and inversion. Continue the traction and maintain the external rotation as you place the hip into extension (touching the table) and adduction. The thigh crosses the midline, and the left side of the trunk elongates. *Caution:* If there is restriction in the range of hip adduction or external rotation the patient's pelvis will move toward the right. If the hip extension is restricted, the pelvis will move into anterior tilt.

Body Mechanics. Stand in a stride position by the patient's left hip with your right foot behind. Face toward the patient's foot and align your body with the line of motion of the pattern. Start with the weight on your front foot and let the motion of the flexing leg push you back over your right leg. If the patient's leg is long, you may have to step back with your left foot as your weight shifts farther back. Continue facing the line of motion.

Alternative Position. You may stand on the right side of the table facing up toward the patient's left hip. Your right hand is on the patient's foot, your left hand on the thigh. Stand in a stride with your right leg forward. As the patient's leg moves up into flexion, step forward with your left leg. If you choose this position, move the patient to the right side of the table (Fig. 8.3 c, d).

Stretch. The reflex comes from a rapid elongation and *rotation* of the hip, ankle, and foot done by both hands simultaneously.

Command. "Foot up, lift your leg up and out." "Lift up!"

Movement. The toes extend as the foot and ankle move into dorsiflexion and eversion. The eversion promotes the hip internal rotation, and these motions occur almost simultaneously. The fifth metatarsal leads as the hip moves into flexion with abduction and internal rotation. Continuation of this motion produces trunk flexion with left side-bending.

Resistance. Your distal hand combines resistance to eversion with traction through the dorsiflexed foot. The resistance to the hip abduction and internal rotation comes from resisting eversion. The traction resists both the dorsiflexion and hip flexion.

Your proximal hand combines traction through the line of the femur with a rotary force that resists the internal rotation and abduction. Maintaining the traction force will guide your resistance in the proper arc. *Caution:* Too much resistance to hip flexion may result in strain on the spine.

End Position. The foot is in dorsiflexion with eversion. The knee is in full extension and the hip in full flexion with enough abduction and internal rotation to align the knee and heel approximately with the lateral border of the left shoulder. *Caution:* The length of the hamstring muscles or other posterior structures may limit the hip motion. Do not allow the pelvis to move into a posterior tilt.

Timing for Emphasis. Prevent motion in the beginning range of hip flexion and exercise the foot and toes.

8.1.3 Flexion-Abduction-Internal Rotation with Knee Flexion (Fig. 8.3 a, b)

Joint	Movement	Muscles: principal components (Kendall and McCreary 1983)
Hip	Flexion, abduction, internal rotation	Tensor fascia lata, rectus femoris, Gluteus medius (anterior), gluteus minimus
Knee	Flexion	Hamstrings, gracilis, gastrocnemius
Ankle/foot	Dorsiflexion, eversion	Peroneus tertius
Toes	Extension, deviation to the left	Extensor hallucis, extensor digitorum

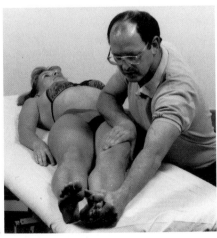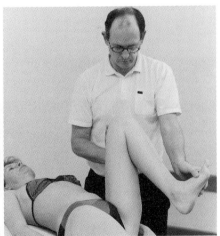

a   b

Fig. 8.3 a, b

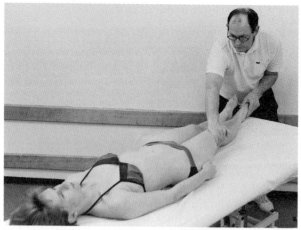

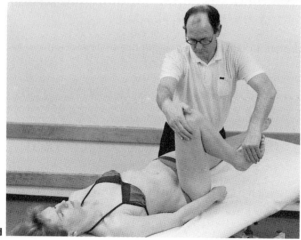

Fig. 8.3 a–d. Leg: flexion – abduction – internal rotation with knee flexion. *a, b* Normal position of the therapist. *c, d* Alternative position on the opposite side of the table

Grip. Your distal and proximal grips remain the same as they were for the straight leg pattern.

Elongated Position. Position the limb as you did for the straight leg pattern.

Body Mechanics. Stand in the same stride position by the patient's hip as for the straight leg pattern. Again, allow the patient to shift your weight from the front to the back foot. Face the line of motion.

Alternative Positions. You may use the same alternative position – standing on the opposite side of the table – as you used for the straight leg pattern (Figs. 8.3 c, d).

Stretch. Use the same motions for the stretch that you used with the straight leg pattern.

Command. "Foot up, bend your knee up and out." "Bend up!"

Movement. The foot and ankle dorsiflex and evert. The hip and knee motions begin next, and both joints reach their end ranges at the same time. Continuation of this motion also causes trunk flexion with lateral flexion to the left.

Resistance. Give traction with your proximal hand through the line of the femur, adding a rotary force, to resist the hip motion. Resist the foot and ankle motion as before with your distal hand. Resist the knee flexion by applying traction through the tibia toward the starting position. The resistance to knee flexion is crucial to successful use of this combination for strengthening the hip and trunk.

End Position. The foot is in dorsiflexion with eversion. The hip and knee are in full flexion with the heel close to the lateral border of the buttock. The knee and heel are aligned with each other and lined up approximately with the lateral border of the left shoulder. If you extend the patient's knee the position is the same as the straight leg pattern.

Note: An anteroposterior plane bisecting the foot should also bisect the knee.

Timing for Emphasis. With three moving segments – hip, knee, and foot – you may lock in any two and exercise the third. With the knee bent it is easy to exercise the internal rotation separately from the other hip motions. Do these exercises where the strength of the hip flexion is greatest. You may work through the full range of hip internal rotation during these exercises, but return to the groove before finishing the pattern.

When exercising the patient's foot, move your proximal hand to a position on the tibia and give resistance to the hip and knee with that hand. Your distal hand is now free to give appropriate resistance to the foot and ankle motions. To avoid fatigue of the hip allow the heel to rest on the table.

8.1.4 Flexion-Abduction-Internal Rotation with Knee Extension (Fig. 8.4)

Joint	Movement	Muscles: principal components (Kendall and McCreary 1983)
Hip	Flexion, abduction, internal rotation	Tensor fascia lata, rectus femoris, gluteus medius (anterior), gluteus minimus
Knee	Extension	Quadriceps
Ankle/foot	Dorsiflexion, eversion	Peroneus tertius
Toes	Extension, deviation to the left	Extensor hallucis, extensor digitorum

Position at Start. For this combination place the patient toward the end of the table so his knee can be flexed as fully as possible.

Grip. Your distal and proximal grips remain the same as they were for the straight leg pattern.

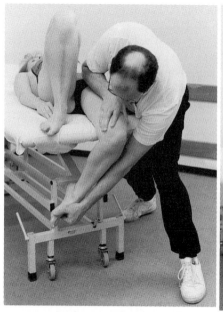
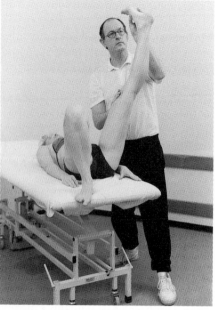
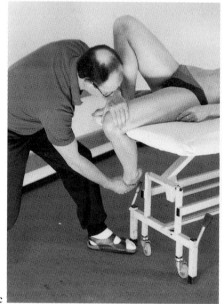

Fig. 8.4 a–d. Leg: flexion – abduction – internal rotation with knee extension.
a, b Normal position of the therapist.
c, d Alternative position at the end of the table

Elongated Position. Traction the entire limb as before while you move the foot into plantar flexion and inversion. Continue the traction and flex the knee over the end of the table as you position the hip in extension with adduction and external rotation. Tightness in the anterior muscles that cross the hip and knee joints may restrict full hip extension-adduction. Keep the thigh in the diagonal and flex the knee

110

as much as possible. *Caution:* Do not allow the pelvis to move to the right or go into anterior tilt.

To protect the patient's back, bend his or her right hip and rest the foot on the end of the table or some other support.

Body Mechanics. Stand in a stride position by the patient's knee. Bend from the hips as you reach down and flex the patient's knee. Your weight shifts back, and then you step back as the patient lifts his leg with the knee extending.

Alternative Positions. Stand at the end of the table facing up towards the patient's left shoulder. Lean back so that your body weight helps with the stretch. As the leg moves into flexion, step forward with your back foot (Fig. 8.4 c, d).

Stretch. Apply the stretch to the hip, knee and foot simultaneously. Stretch the *hip* with the proximal hand, using rapid elongation and rotation. Stretch the *ankle and foot* with your distal hand, using traction and rotation. Stretch the *knee* very gently by applying traction with your distal hand along the line of the tibia.

Command. "Foot up, bend your hip up and straighten your knee as you go."

Movement. The foot and ankle dorsiflex and invert. The hip motion begins next. When the tip has moved through about 5° of flexion the knee begins to extend. It is important that the hip and knee reach their end ranges at the same time.

Resistance. Your distal hand resists the foot and ankle motion with a rotary push. Using the stable foot as a handle, resist the knee extension with a traction force toward the starting position of knee flexion. The rotary resistance at the foot resists the knee and hip rotation as well.

Your proximal hand combines traction through the line of the femur with a twist to resist internal rotation. *Note:* The knee takes more resistance than the hip. Your two hands must work separately.

End Position. The end position is the same as the straight leg pattern.

Timing for Emphasis. The emphasis here is to teach the patient to combine hip flexion with knee extension in a smooth motion.

8.1.5 Extension-Adduction-External Rotation (Fig. 8.5)

Joint	Movement	Muscles: principal components (Kendall and McCreary 1983)
Hip	Extension, adduction, external rotation	Adductor magnus, gluteus maximus, Hamstrings, lateral rotators
Knee	Extended (position unchanged)	Quadriceps
Ankle/foot	Plantar flexion, inversion	Gastrocnemius, soleus, tibialis posterior
Toes	Flexion, deviation to the right	Flexor hallucis, flexor digitorum

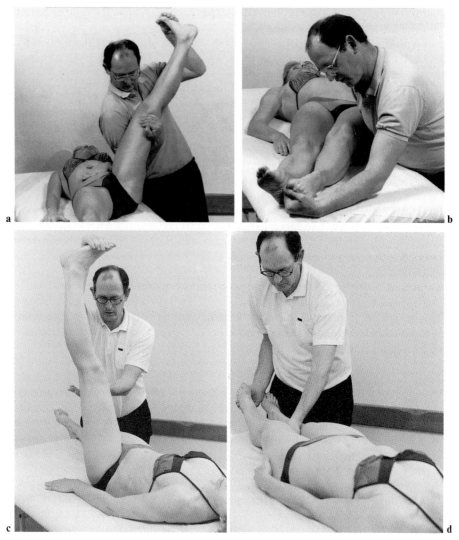

Fig. 8.5 a–d. Leg: extension – adduction – external rotation. ***a, b*** Normal position of the therapist. ***c, d*** Alternative position on the other side of the table

Grip. *Distal hand:* Hold the plantar surface of the foot with the palm of your left hand. Your thumb is at the base of the toes to facilitate toe flexion. Your fingers hold the medial border of the foot, the heel of your hand gives counterpressure along the lateral border. *Caution:* Do not squeeze or pinch the foot.

Proximal hand: Your right hand comes underneath the thigh from lateral to medial to hold on the posteromedial side.

112

Elongated Position. Traction the entire leg while moving the foot into dorsiflexion and eversion. Continue the traction and maintain the internal rotation as you lift the leg into flexion and abduction. If the patient has just completed the antagonistic motion (flexion-abduction-internal rotation), begin at the end of that pattern. *Caution:* Do not try to push the hip past the limitation imposed by hamstring length. Do not allow the pelvis to move into a posterior tilt.

Body Mechanics. Stand in a stride position by the patient's left shoulder facing toward the lower right corner of the table. Your inner foot (left) is in front. Your weight is on the back foot (right). Allow the patient to pull you forward onto your front foot. When your weight has shifted over the front foot, step forward with your rear foot and continue the weight shift forward.

Alternative Position. You may stand on the right side of the table facing up toward the left hip. Your right hand is on the plantar surface of the patient's foot, your left hand on the posterior thigh. Stand in a stride and allow the patient to push your weight back as the leg kicks down (Fig. 8.5 c, d).

Stretch. Your proximal hand stretches the hip by giving a quick traction to the thigh. Use the forearm of your distal hand to traction up through the shin while you stretch the patient's foot farther into dorsiflexion and eversion. *Caution:* Do not force the hip into more flexion.

Command. "Point your toes, push your foot down and kick down and in." "Push!"

Movement. The toes flex and the foot and ankle plantar flex and invert. The inversion promotes the hip external rotation, and these motions occur at the same time. The fifth metatarsal leads as the thigh moves down into extension and adduction maintaining the external rotation. Continuation of this motion causes extension with elongation of the left side of the trunk.

Resistance. Your distal hand combines resistance to inversion with approximation through the bottom of the foot. The approximation resists both the plantar flexion and the hip extension. Resisting inversion results in resistance to the hip adduction and external rotation as well. Your proximal hand lifts the thigh back towards the starting position. The lift resists the hip extension and adduction. The placement of your hand, coming from lateral to medial, gives resistance to the external rotation.

As the hip approaches full extension, continue to give approximation through the foot with your distal hand and approximate through the thigh with your proximal hand.

End Position. The foot is in plantar flexion with inversion and the toes are flexed. The knee remains in full extension. The hip is in extension (touching the table) and adduction while maintaining external rotation. The thigh has crossed to the right side of the midline.

Timing for Emphasis. Lock in the hip at the end of the range and exercise the foot and toes.

113

8.1.6 Extension-Adduction-External Rotation with Knee Extension
(Fig. 8.6)

Joint	Movement	Muscles: principal components (Kendall and McCreary 1983)
Hip	Extension, adduction, external rotation	Adductor magnus, gluteus maximus, hamstrings, lateral rotators
Knee	Extension	Quadriceps
Ankle/foot	Plantar flexion, inversion	Gastrocnemius, soleus, tibialis posterior
Toes	Flexion, medial deviation	Flexor hallucis, flexor digitorum

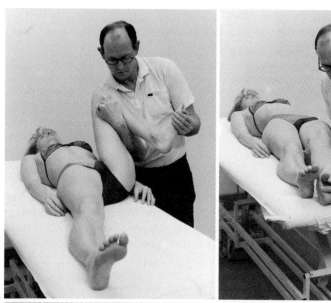
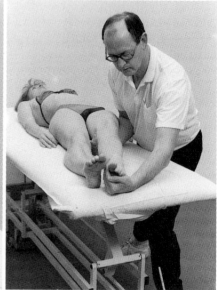
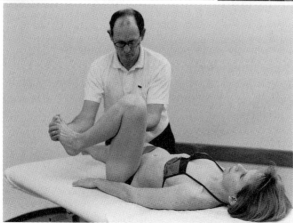

Fig. 8.6 a–d. Leg: extension – adduction – external rotation with knee extension. *a, b* Normal position of the therapist. *c, d* Alternative position on the other side in the groove

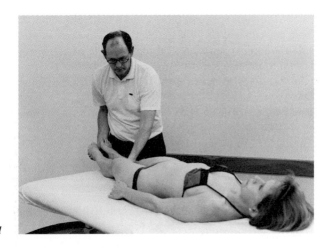

Fig. 8.6 d

Grip. Your distal and proximal grips are the same as the ones used for the straight leg pattern.

Elongated Position. The foot is in dorsiflexion with eversion. The hip and knee are in full flexion with the heel close to the lateral border of the buttock. The knee and heel are aligned with each other and lined up approximately with the lateral border of the left shoulder. The hip has the same amount of rotation as it did in the straight leg pattern. Straighten the knee to check the rotation.

Body Mechanics. Your body mechanics are the same as for the straight leg pattern.

Alternative Position. You may stand on the opposite side of the table facing up toward the left hip (Fig. 8.6 c, d).

Stretch. Apply the stretch to the hip, knee, and foot simultaneously. With your *proximal* (right) hand combine traction of the hip through the line of the femur with a rotary motion to stretch the external rotation. Your *distal* (left) hand stretches the foot farther into dorsiflexion and eversion and stretches the knee extension by bringing the patient's heel closer to his buttock.

Command. "Push your foot down and kick down and in." "Kick!"

Movement. The foot and ankle plantar flex and invert. The hip motion begins next. When the hip extension has completed about 5° of motion the knee begins to extend. It is important that the hip and knee reach their end ranges at the same time.

Resistance. Your distal hand resists the foot and ankle motion with a rotary push. Using the foot as a handle, resist the knee extension by pushing the patient's heel back toward the buttock. The angle of this resistance will change as the knee moves further into extension. The rotary resistance at the foot also resists the rotation at the knee and hip. The resistance to the knee extension motion continues in the

115

same direction (toward the patient's buttock) when the knee is fully extended. Your proximal hand pulls the thigh back toward the starting position. The pull resists the hip extension and adduction. The placement of your hand, coming from lateral to medial, supplies the resistance to the external rotation. *Note:* The knee takes more resistance than the hip. Your two hands must work separately.

End Position. The end position is the same as the straight leg pattern.

Timing for Emphasis. Prevent knee extension at the beginning of the range and exercise the hip motions. Lock in hip extension in mid-range and exercise the knee extension.

8.1.7 Extension-Adduction-External Rotation with Knee Flexion (Fig. 8.7)

Joint	Movement	Muscles: principal components (Kendall and McCreary 1983)
Hip	Extension, adduction, external rotation	Adductor magnus, gluteus maximus, lateral rotators
Knee	Flexion	Hamstrings, gracilis
Ankle/foot	Plantar flexion, inversion	Soleus, tibialis posterior
Toes	Flexion, medial deviation	Flexor hallucis, flexor digitorum

Position at Start. For this combination place the patient toward the end of the table so that the knee can flex as fully as possible. This is the same position as you used to begin the pattern of flexion-abduction-internal rotation with knee extension (Sect. 8.1.4). You may to protect the patient's back by bending his or her right hip and resting the foot on the end of the table or some other support.

Grip. Your distal and proximal grips are the same as those used for the straight leg pattern.

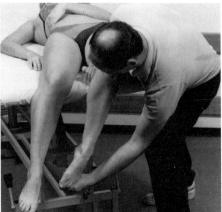

a   b

Fig. 8.7 a, b. Leg: extension – adduction – external rotation with knee flexion

Elongated Position. Position the limb as you did for the straight leg pattern.

Body Mechanics. Use the same body mechanics as for the straight leg pattern. As the pattern nears end range, bend at your hips as you reach down to continue resisting the knee flexion.

Stretch. The stretch comes from the rapid elongation and rotation of the hip, ankle, and foot by both hands simultaneously.

Command. "Push your foot and toes down; push your hip down and bend your knee as you go."

Movement. The foot and ankle plantar flex and invert. The hip motion begins next. When the hip extension has completed about 5° of motion the knee begins to flex. It is important that the hip and knee reach their end ranges at the same time.

Resistance. Your distal hand uses the resistance to the plantar flexion and inversion to resist the knee flexion as well. The pull is back towards the starting position of knee extension and foot eversion. Your proximal hand resists the hip motion as it did for the straight leg pattern.

End Position. The hip is extended with adduction and external rotation. The knee is flexed over the end of the table and the foot is in plantar flexion with inversion. *Caution:* Do not allow the pelvis to move to the right or go into anterior tilt.

Timing for Emphasis. Lock in the hip extension at any point in the range and exercise the knee flexion. Do not let the hip action change from extension to flexion. Teach the patient to combine hip extension with knee flexion in a smooth motion.

8.1.8 Flexion-Adduction-External Rotation (Fig. 8.8)

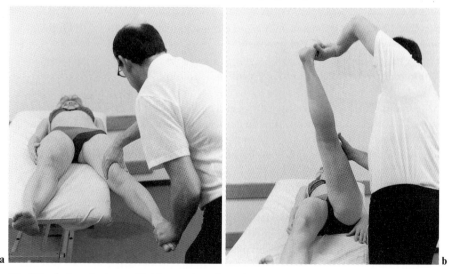

Fig. 8.8 a, b. Leg: flexion – adduction – external rotation

Joint	Movement	Muscles: principal components (Kendall and McCreary 1983)
Hip	Flexion, adduction, external rotation	Psoas major, iliacus, adductor muscles, Sartorius, pectineus, rectus femoris
Knee	Extended (position unchanged)	Quadriceps
Ankle/foot	Dorsiflexion, inversion	Tibialis anterior
Toes	Extension, medial deviation	Extensor hallucis, extensor digitorum

Grip. *Distal hand:* Your left hand grips the patient's foot with the fingers on the medial border and the thumb giving counterpressure on the lateral border. Hold the sides of the foot but do not put any contact on the plantar surface. To avoid blocking toe motion, keep your grip proximal to the metatarsal-phalangeal joints. *Caution:* Do not squeeze or pinch the foot.

Proximal hand: Place your right hand on the anterior-medial surface of the thigh just proximal to the knee.

Elongated Position. Traction the entire limb while you move the foot into plantar flexion and eversion. Continue the traction and maintain the internal rotation as you place the hip into hyperextension and abduction. The trunk elongates diagonally from right to left. *Caution:* If the hip extension is restricted, the pelvis will move into anterior tilt. If the abduction is restricted, the pelvis will move to the left.

Body Mechanics. Stand in a stride position by the patient's left foot with your left foot behind. Face toward the patient's right shoulder with your body aligned with the pattern's line of motion. Shift your weight from your front foot to your back foot as you stretch. As the patient moves, let the resistance shift your weight forward over your front foot. If the patient's leg is long, you may have to take a step as your weight shifts farther forward. Continue facing the line of motion.

Stretch. The reflex comes from a rapid elongation and rotation of the hip, ankle, and foot by both hands simultaneously.

Command. "Foot up, lift your leg up and in." "Lift up!"

Movement. The toes extend as the foot and ankle move into dorsiflexion and inversion. The inversion promotes the hip external rotation, so these motions occur simultaneously. The big toe leads as the hip moves into flexion with adduction and external rotation. Continuation of this motion produces trunk flexion to the right.

Resistance. Your distal hand combines resistance to inversion with traction through the dorsiflexed foot. The resistance to the hip adduction and external rotation comes from resisting the inversion. The traction resists both the dorsiflexion and hip flexion. Your proximal hand combines traction through the line of the femur with a rotary force to resist the external rotation and adduction. Maintaining the traction force will guide your resistance in the proper arc. *Caution:* Too much resistance to hip flexion may result in strain on the patient's spine.

118

End Position. The foot is in dorsiflexion with inversion. The knee is in full extension. The hip is in full flexion with enough adduction and external rotation to place the knee and heel in a diagonal line with the right shoulder. *Caution:* The length of the hamstring muscles or other posterior structures may limit the hip motion. Do not allow the pelvis to move into a posterior tilt.

Timing for Emphasis. You may prevent motion in the beginning range of hip flexion and exercise the foot and toes.

8.1.9 Flexion-Adduction-External Rotation with Knee Flexion (Fig. 8.9)

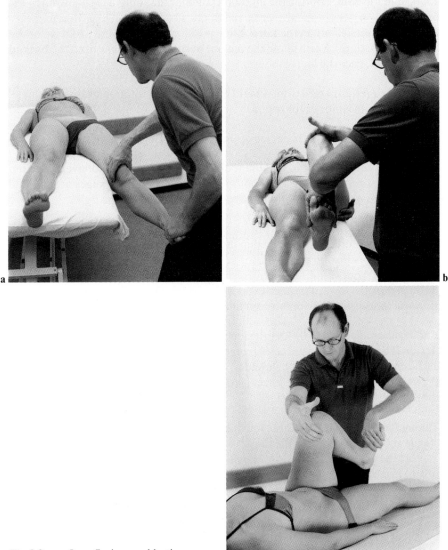

Fig. 8.9 a–c. Leg: flexion – adduction – external rotation with knee flexion

119

Joint	Movement	Muscles: principal components (Kendall and McCreary)
Hip	Flexion, adduction, external rotation	Psoas major, iliacus, adductor muscles, sartorius, pectineus, rectus femoris
Knee	Flexion	Hamstrings, gracilis, gastrocnemius
Ankle/foot	Dorsiflexion, inversion	Tibialis anterior
Toes	Extension, medial deviation	Extensor hallucis, extensor digitorum

Grip. Your grips are the same as those for the straight leg pattern.

Elongated Position. Position the limb as you did for the straight leg pattern.

Body Mechanics. Stand in the same stride position by the patient's foot as for the straight leg pattern. Again allow the patient to shift your weight from the back to the front foot. Face the line of motion.

Stretch. The reflex comes from a rapid elongation and rotation of the hip, ankle, and foot by both hands simultaneously.

Command. "Foot up, bend your leg up and across." "Bend up!"

Movement. The toes extend and the foot and ankle dorsiflex and invert. The hip and knee flexion begin next, and both joints reach their end ranges at the same time. Continuation of this motion also causes trunk flexion to the right.

Resistance. Give traction with your proximal hand through the line of the femur, adding a rotary force, to resist the hip motion. The resistance given by your distal hand to the dorsiflexion and inversion will also resist the hip adduction and external rotation. Your distal hand now resists the knee flexion by applying traction through the tibia toward the starting position. *Note:* The resistance to knee flexion is crucial to successful use of this combination for strengthening the hip and trunk.

End Position. The foot is in dorsiflexion with inversion, the hip and knee are in full flexion. The adduction and external rotation cause the heel and knee to line up with each other and with the right shoulder. *Note:* An anteroposterior plane bisecting the foot should also bisect the knee. If you extend the patient's knee, the position is the same as the straight leg pattern.

Timing for Emphasis. With three moving segments – hip, knee and foot – you may lock in any two and exercise the third.

With the knee bent it is easy to exercise the external rotation separately from the other hip motions. Do these exercises where the strength of the hip flexion is greatest. You may work through the full range of hip external rotation during these exercises. Return to the groove before finishing the pattern.

When exercising the foot, move your proximal hand to a position on the tibia and give resistance to the hip and knee with that hand. Your distal hand is now free to give appropriate resistance to the foot and ankle motions. To avoid fatigue of the hip allow the heel to rest on the table.

8.1.10 Flexion-Adduction-External Rotation with Knee Extension
(Fig. 8.10)

Joint	Movement	Muscles: principal components (Kendall and McCreary 1983)
Hip	Flexion, adduction, external rotation	Psoas major, iliacus, adductor muscles, sartorius, pectineus, rectus femoris
Knee	Extension	Quadriceps
Ankle/foot	Dorsiflexion, inversion	Tibialis anterior
Toes	Extension, medial deviation	Extensor hallucis, extensor digitorum

Grip. Your grips remain the same as those for the straight leg pattern.

Elongated Position. Traction the entire limb as before while you move the foot into plantar flexion and eversion. Continue the traction and flex the knee over the side of the table as you position the hip in extension with abduction and internal rotation. Tightness in the anterior muscles that cross the hip and knee joints may restrict full hip extension-abduction. Keep the thigh in the diagonal and flex the knee as much as possible. *Caution:* Do not allow the pelvis to move to go into anterior tilt. To protect the patient's back bend his or her right hip and rest the foot on the table.

Body Mechanics. Stand in a stride position by the patient's knee facing the foot of the table. Bend from the hips to reach down and flex the patient's knee. Your weight shifts forward, and then you turn to face the line of the pattern. Step forward as the patient lifts his leg with the knee extending.

Stretch. Apply the stretch to the hip, knee, and foot simultaneously. Stretch the *hip* with the proximal hand using rapid elongation and rotation. Stretch the *ankle and*

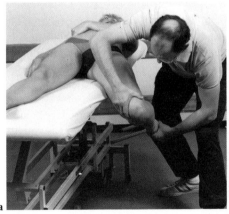

Fig. 8.10 a, b. Leg: flexion – adduction – external rotation with knee extension

121

foot with your distal hand using traction and rotation. Stretch the *knee* very gently by applying traction with your distal hand along the line of the tibia.

Command. "Foot up, bend your hip up and straighten your knee as you go."

Movement. The foot and ankle dorsiflex and evert. The hip motion begins next. When the hip has moved through about 5° of flexion the knee begins to extend. It is important that the hip and knee reach their end ranges at the same time.

Resistance. Your distal hand resists the foot and ankle motion with a rotary push. Using the stable foot as a handle, resist the knee extension with a traction force toward the starting position of knee flexion. The rotary resistance at the foot resists the knee and hip rotation as well.

Your proximal hand combines traction through the line of the femur with a twist to resist the external rotation and adduction. *Note:* The knee takes more resistance than the hip. Your two hands must work separately.

End Position. The end position is the same as the straight leg pattern.

Timing for Emphasis. The emphasis here is to teach the patient to combine hip flexion with knee extension in a smooth motion.

8.1.11 Extension-Abduction-Internal Rotation (Fig. 8.11)

Joint	Movement	Muscles: principal components (Kendall and McCreary 1983)
Hip	Extension, abduction, internal rotation	Gluteus medius, gluteus maximus (upper), hamstrings
Knee	Extended (position unchanged)	Quadriceps
Ankle/foot	Plantar flexion, eversion	Gastrocnemius, soleus, peroneus longus and brevis
Toes	Flexion, lateral deviation	Flexor hallucis, flexor digitorum

Grip. *Distal hand:* Hold the foot with the palm of your left hand along the plantar surface. Your thumb is at the base of the toes to facilitate toe flexion. Your fingers hold the medial border of the foot while the heel of your hand gives counter pressure along the lateral border. *Caution:* Do not squeeze or pinch the foot.

Proximal hand: Your right hand holds the posterolateral side of the thigh.

Elongated Position. Traction the entire leg while moving the foot into dorsiflexion and inversion. Continue the traction and maintain the internal rotation as you lift the leg into flexion and adduction. *Caution:* Do not try to push the hip past the limitation imposed by hamstring length. Do not allow the pelvis to move into a posterior tilt.

If the patient has just completed the antagonistic motion (flexion-adduction-external rotation), begin at the end of that pattern.

Body Mechanics. Stand in a stride position facing the patient's right shoulder. Your weight is on the front foot. Allow the patient to push your back onto your rear foot,

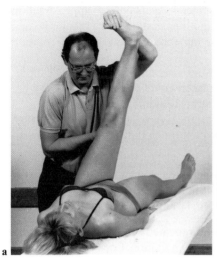
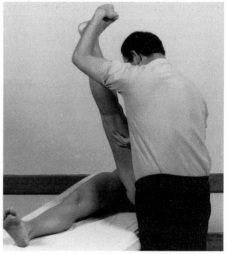
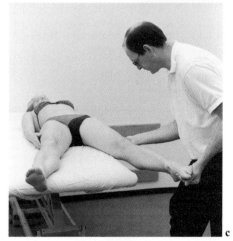

Fig. 8.11 a–c. Leg: extension – abduction – internal rotation

then step back and continue to shift your weight backward. Keep your elbows close to your sides so you can give the resistance with your body and legs.

Stretch. The proximal hand gives a stretch by rapid traction of the thigh.

Use the forearm of your distal hand to traction up through the shin while you stretch the patient's foot farther into dorsiflexion and inversion. *Caution:* Do not force the hip into more flexion.

Command. "Point your toes, push your foot down and kick down and out." "Push!"

Movement. The toes flex and the foot and ankle plantar flex and evert. The eversion promotes the hip internal rotation; these motions occur at the same time. The thigh moves down into extension and abduction, maintaining the internal rotation. Continuation of this motion causes extension with left side bending of the trunk.

123

Resistance. Your distal hand combines resistance to eversion with approximation through the bottom of the foot. The approximation resists both the plantar flexion and the hip extension. The resistance to the hip abduction and internal rotation comes from the resisted eversion.

Your proximal hand lifts the thigh back towards the starting position. The lift resists the hip extension and abduction. The placement of your hand, coming from lateral to posterior, gives resistance to the internal rotation.

As the hip approaches full extension, continue to give approximation through the foot with your distal hand and approximate through the thigh with your proximal hand.

End Position. The foot is in plantar flexion with inversion and the toes are flexed. The knee remains in full extension. The hip is in as much hyperextension as possible while maintaining the abduction and internal rotation.

Timing for Emphasis. Use approximation with repeated contractions or combination of isotonics to exercise the hyperextension hip motion. Lock in the hip at the end of the range and exercise the foot and toes.

8.1.12 Extension-Abduction-Internal Rotation with Knee Extension
(Fig. 8.12)

Joint	Movement	Muscles: principal components (Kendall and McCreary 1983)
Hip	Extension, abduction, internal rotation	Gluteus medius, gluteus maximus (upper), Hamstrings
Knee	Extension	Quadriceps
Ankle/foot	Plantar flexion, eversion	Gastrocnemius, soleus, peroneus longus and brevis,
Toes	Flexion, lateral deviation	Flexor hallucis, flexor digitorum

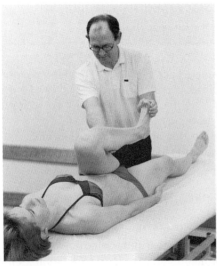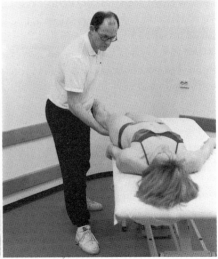

Fig. 8.12 a, b. Leg: extension – abduction – internal rotation with knee extension

124

Grip. Your grips are the same as for the straight leg pattern.

Elongated Position. The foot is in dorsiflexion with inversion. The hip and knee are in full flexion with the heel close to the right buttock. The knee and heel are aligned with the each other and lined up approximately with the right shoulder.

Body Mechanics. Your body mechanics are the same as for the straight leg pattern.

Stretch. Apply the stretch to the hip, knee, and foot simultaneously. With your *proximal hand* combine traction of the hip through the line of the femur with a rotary motion to stretch the internal rotation. Your *distal hand* stretches the foot farther into dorsiflexion and inversion as you stretch the knee extension by bringing the patient's heel closer to his or her buttock.

Command. "Push your foot down and kick down and out." "Kick!"

Movement. The foot and ankle plantar flex and evert. The hip motion begins next. When the hip extension has completed about 5° of motion the knee begins to extend. It is important that the hip and knee reach their end ranges at the same time.

Resistance. Your distal hand resists the foot and ankle motion with a rotary push. Using the foot as a handle, resist the knee extension by pushing the patient's heel back toward the buttock. The angle of this resistance will change as the knee moves further into extension. The rotary resistance at the foot resists the knee and hip rotation as well. *Note:* The resistance to the knee extension motion continues in the same direction when the knee is full extended.

Your proximal hand lifts the thigh back towards the starting position. The lift resists the hip extension and abduction. The placement of the hand from lateral to posterior gives resistance to the internal rotation. *Note:* The knee takes more resistance than the hip. Your two hands must work separately.

End Position. The end position is the same as the straight leg pattern.

Timing for Emphasis. Prevent knee extension at the beginning of the range and exercise the hip motions. Lock in hip extension in mid-range and exercise knee extension.

8.1.13 Extension-Abduction-Internal Rotation with Knee Flexion (Fig. 8.13)

Joint	Movement	Muscles: principal components (Kendall and McCreary 1983)
Hip	Extension, abduction, internal rotation	Gluteus medius, gluteus maximus (upper)
Knee	Flexion	Hamstrings, gracilis
Ankle/foot	Plantar flexion, eversion	Soleus, peroneus longus and brevis
Toes	Flexion, lateral deviation	Flexor hallucis, flexor digitorum

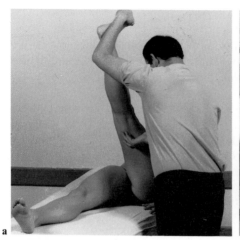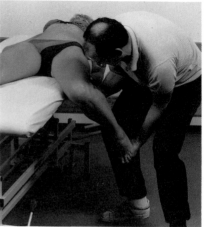

a   b

Fig. 8.13 a, b. Leg: extension – abduction – internal rotation with knee flexion

Position at Start. You may wish to protect the patient's back by bending the right hip and resting the foot on the table.

Grip. Your grips are the same as those for the straight leg pattern.

Elongated Position. Position the limb as you did for the straight leg pattern.

Body Mechanics. Use the same body mechanics as for the straight leg pattern. As the pattern nears end range, bend at your hips as you reach down to continue resisting the knee flexion. You may turn your body to face toward the foot of the table as the knee and hip reach their end range.

Stretch. The reflex comes from the rapid elongation and rotation of the hip, ankle, and foot by both hands simultaneously.

Command. "Push your foot and toes down, push your hip down, and bend your knee as you go."

Movement. The foot and ankle plantar flex and evert. The hip motion begins next. When the hip extension has completed about 5° of motion the knee begins to flex. It is important that the hip and knee reach their end ranges at the same time.

Resistance. Your distal hand resists the plantar flexion and eversion and uses that force to resist the knee flexion as well. The force is back towards the starting position of knee extension and foot inversion. Your proximal hand resists the hip motion as it did for the straight leg pattern.

End Position. The hip is extended with abduction and internal rotation. The knee is flexed over the side of the table and the foot is in plantar flexion with eversion. *Caution:* Do not allow the pelvis to got into anterior tilt.

Timing for Emphasis. Lock in the hip extension at any point in the range and exercise the knee flexion. Do not let the hip action change from extension to flexion. Teach the patient to combine hip extension with knee flexion in a smooth motion.

8.2 Changing the Patient's Position

There are many advantages to exercising the patient in a variety of positions. These include the patient's ability to see his or her leg, adding or eliminating the effect of gravity from a motion, and putting two-joint muscles on stretch. There are also disadvantages for each position. Choose the positions that give the most advantages with the fewest drawbacks.

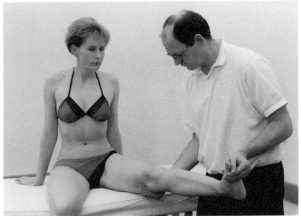

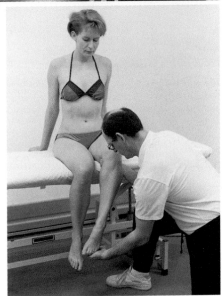

Fig. 8.14 a, b

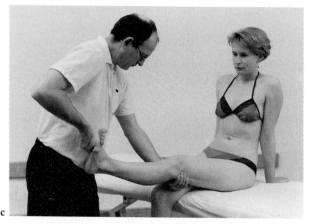

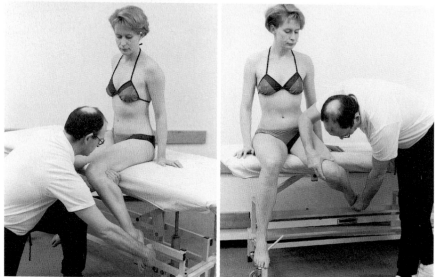

Fig. 8.14 a–f. Leg patterns in a sitting position. ***a, b*** Extension – adduction – external rotation with knee flexion. ***c, d*** Extension – abduction – internal rotation with knee flexion. ***e, f*** Flexion – adduction – external rotation with knee extension

8.2.1 Leg Patterns in a Sitting Position

The sitting position allows the therapist to work with the legs when hip extension is restricted by an outside force. This position lets the patient see the foot and knee while they are exercising. In addition, working in this position challenges the patient's sitting balance and stability. The number of lower extremity exercises that you can do with your patients in sitting is limited only by the patient's abilities and your imagination. We have pictured three examples in Fig. 8.14.

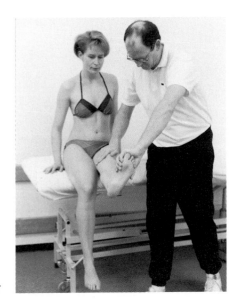

Fig. 8.14 f

8.2.2 Leg Patterns in a Prone Position (Fig. 8.15)

Working with the patient in a prone position allows you to exercise the hip hyperextension against gravity. This can be good position in which to exercise the combination of hip hyperextension with knee flexion. *Note:* Be careful to restrict the motion to the hip. Do not allow the lumbar spine to hyperextend.

To exercise hip flexion in a prone position the patient must be positioned with the legs over the end of the table.

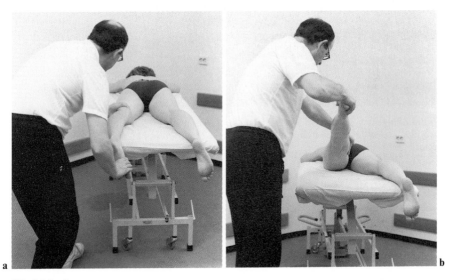

Fig. 8.15 a, b

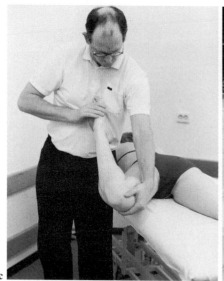
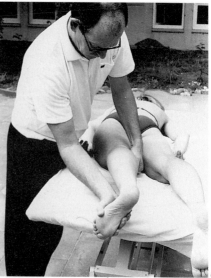

c

d

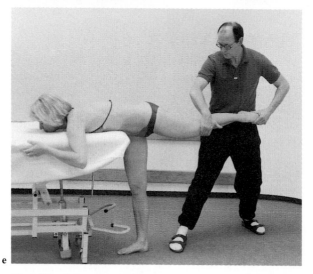

e

Fig. 8.15 a–f. Leg patterns in a prone position. *a, b* Extension – adduction – external rotation. *c, d* Flexion – adduction – external rotation with knee extension. *e, f* Flexion – abduction – internal rotation with knee flexion

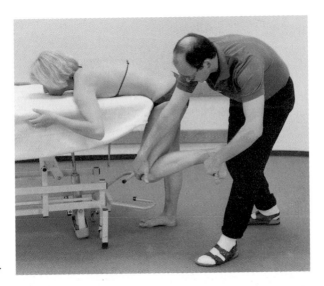

Fig. 8.15f

8.2.3 Leg Patterns in a Sidelying Position (Fig. 8.16)

When working with the patient in sidelying, take care that the patient does not substitute trunk motion or pelvic rolling for the leg motions you want to exercise. You may stabilize the patient's trunk with external support or let the patient do the work of stabilizing the trunk him- or herself. In this position the abductor muscles of the upper leg and the adductor muscles of the lower leg work against gravity. This position is useful for exercising hip hyperextension. Use approximation and resistance to rotation to facilitate the motion.

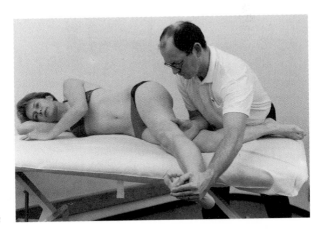

Fig. 8.16a

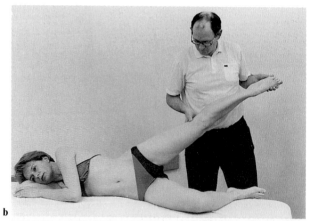

b

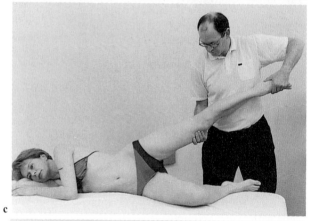

c

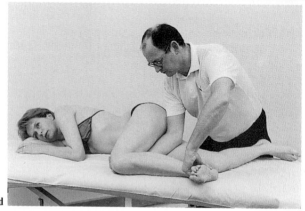

d

Fig. 8.16 b–d

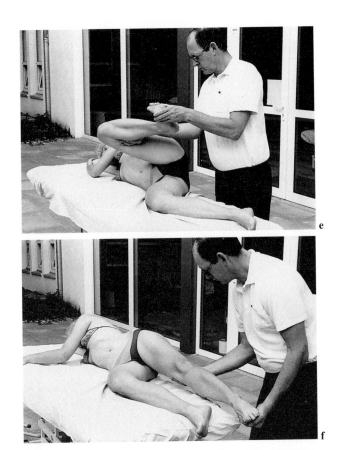

Fig. 8.16 a–f. Leg patterns in a sidelying position. **a, b** Extension – abduction – internal rotation with a straight knee. **c, d** Flexion – adduction – external rotation with knee flexion. **e, f** Extension – adduction – external rotation with knee extension

8.2.4 Leg Patterns in a Crawling Position (Fig. 8.17)

Working in this position requires that the patient stabilize his or her trunk and bear weight on the arms as well as on the nonmoving leg. As in the prone position, the hip extensor muscles work against gravity. The hip flexion can move through its full range with gravity eliminated. *Caution:* Do not allow the spine to move into undesired positions or postures.

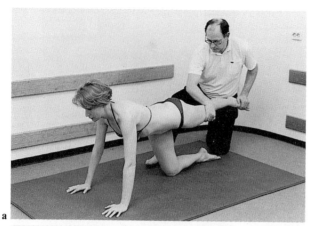

a

b

c

Fig. 8.17 a–c

d

e

f

Fig. 8.17 d–f

135

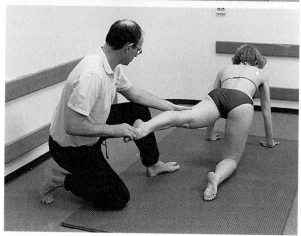

Fig. 8.17 a–h. Leg patterns in a crawling position. *a, b* Flexion – abduction – internal rotation with knee flexion. *c, d* Extension – adduction – external rotation with knee extension. *e, f* Flexion – adduction – external rotation with knee flexion. *g, h* Extension – abduction – internal rotation with knee extension

8.2.5 Bilateral Leg Patterns

When you exercise both legs at the same time there is always more demand on the trunk muscles than when only one leg is exercising. To exercise the trunk specifically you hold both the legs together. The leg patterns for trunk exercise are discussed in Chap. 10.

When you hold the legs separately the emphasis of the exercise is on the legs. Bilateral leg work allows you to use irradiation from the patient's strong leg to facilitate weak motions or muscles in the involved leg. You can use any combination of patterns in any position. Work with those that give you and the patient the greatest advantage in strength and control.

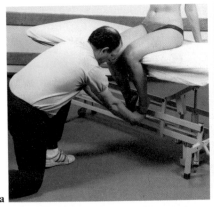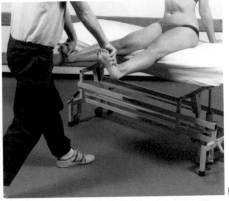

Fig. 8.18 a, b. Leg: bilateral symmetrical pattern of flexion – abduction with knee extension

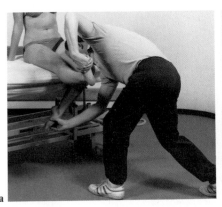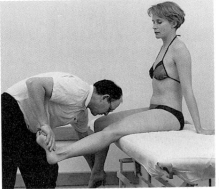

Fig. 8.19 a, b. Leg: bilateral asymmetrical pattern of flexion – abduction with knee extension on the left, extension – abduction with knee flexion on the right

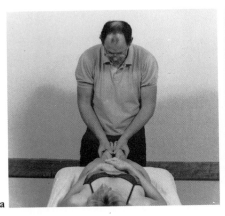

Fig. 8.20 a, b

137

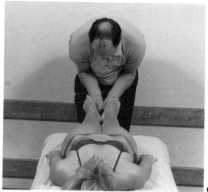

c d

Fig. 8.20 a–d. Leg: symmetrical straight leg combinations in a supine position. **a, b** Flexion – abduction, **c, d** extension – adduction

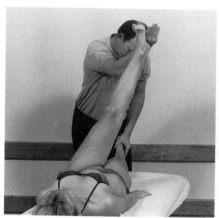

a b

Fig. 8.21 a, b. Leg: bilateral asymmetrical reciprocal pattern of extension – abduction on the left, flexion – abduction on the right

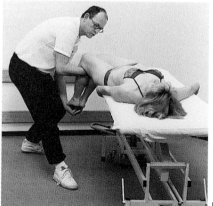

a b

Fig. 8.22 a, b. Leg: bilateral asymmetrical pattern of hip extension with knee flexion: left leg in abduction and right leg in adduction

138

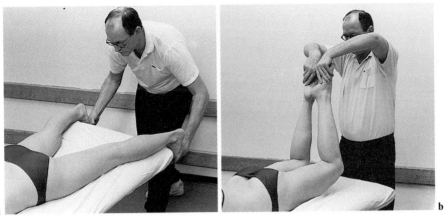

Fig. 8.23 a, b. Leg: bilateral symmetrical pattern in a prone position: hip extension adduction, and external rotation with knee flexion

The most common positions for doing bilateral leg patterns are supine, prone, and sitting. In sitting we show two possible combinations. The first is a bilateral symmetrical combination, flexion-abduction with knee extension (Fig. 8.18), and the second is a reciprocal asymmetrical combination, left leg flexion-abduction with knee extension combined with right leg extension-abduction with knee flexion (Fig. 8.19). In a supine position, the symmetrical straight leg combinations of flexion-abduction (Fig. 8.20 a, b) and extension-adduction (Fig. 8.20 c, d), the reciprocal combination of left leg extension-abduction with right leg flexion-abduction (Fig. 8.21), and the asymmetrical pattern of hyperextension with kneeflexion (Fig. 8.22) are shown. In the prone position we show hip extension with knee flexion (Fig. 8.23).

Reference

Kendall FP, McCreary EK (1983) Muscles, testing and function. Williams and Wilkins, Baltimore

9 The Neck

9.1 Introduction

There are many reasons to exercise the neck patterns. Movement of the head and neck helps to guide trunk motions. Resistance to neck motion provides irradiation for trunk muscle exercises. You can use the neck patterns when you want to treat dysfunctions in the cervical and thoracic spine directly. Finally, stability of the head and neck are essential for most everyday activities.

In this chapter we cover the basic neck patterns and the use of the neck for facilitation of trunk motions.

9.1.1 Diagonal Motion

The neck patterns include the same three motion components as the other patterns: flexion or extension, lateral flexion, and rotation. A plane through the nose, the chin, and the crown of the head defines the proper course of the pattern.

The distal component in the neck patterns is the upper cervical spine. The motion is called *short* neck flexion or *short* neck extension. The proximal component is the lower cervical spine and upper thoracic spine to T6. We call the motion here *long* neck flexion or *long* neck extension.

Movements of the head and eyes reinforce each other. The range of neck motion will be limited if the patient does not look in the direction of the head movement. Giving the patient a specific spot to look at guides the neck motion. Conversely, movement of the head in the appropriate direction facilitates eye motions.

Jaw motion is associated with movement of the head on the neck. Mouth opening and short neck flexion reinforce each other. Mouth closing and short neck extension reinforce each other.

Continuation of the neck flexion patterns results in trunk flexion. Continuation of the neck extension pattern results in trunk elongation. The neck rotation pattern facilitates trunk lateral flexion.

The neck flexion-extension diagonals are

1. Flexion-right lateral flexion-right rotation and extension-left lateral flexion-left rotation.
2. Flexion-left lateral flexion-left rotation and extension-right lateral flexion-right rotation.

The neck rotation pattern is:

1. Full neck rotation
2. Ipsilateral lateral flexion
3. Short neck flexion
4. Long neck extension

In this chapter we illustrate and describe the diagonal of flexion to the left, extension to the right. To work with the other diagonal, reverse the words "left" and "right" in the instructions. We suggest that you have your patient sitting when you begin exercising the neck patterns.

Neck Flexion-Lateral Flexion-Rotation

Movement	Muscles: principal components (Kendall and McCreary 1983)
Short neck flexion	Longus capitis, rectus capitis anterior, suprahyoid muscles (tuck chin), Infrahyoid muscles (stabilize hyoid)
Long neck flexion	Longus colli, platysma, scalenus anterior, sternocleidomastoid
Rotation	*Contralateral:* Scalenus (all), sternocleidomastoid *Ipsilateral:* Longus capitis and colli, Rectus capitis anterior
Lateral flexion	Longus colli, scalenus (all), sternocleidomastoid

Neck Extension-Lateral Flexion-Rotation

Extension	Muscles: principal components (Kendall and McCreary 1983)
Short neck extension	Iliocostalis and longissimus capitis, obliquus capitis (superior and inferior), rectus capitis posterior (major and minor), semispinalis and splenius capitis, trapezius,
Long neck extension	Iliocostalis cervicis, longissimus and splenius cervicis, multifidi and rotatores, semispinalis and splenius cervicis, trapezius
Rotation	*Contralateral:* Multifidi and rotatores, semispinalis capitis, upper trapezius *Ipsilateral:* Obliquus capitis inferior, splenius cervicis and capitis,
Lateral flexion	Iliocostalis cervicis, intertransversarii (cervical), longissimus capitis, obliquus capitis superior, splenius cervicis and capitis, trapezius

9.1.2 Patient Position

Sitting is a functional position for neck motion and stability. In the prone on elbows position the neck extensor muscles must work against gravity while neck flexion has gravity assistance. In a supine position, neck flexion will assist the patient in rolling and getting to sitting. However, the flexor muscles must be strong enough to lift the head against gravity. Sidelying eliminates the effects of gravity from the motions of flexion and extension. In this position it is easy to use resisted neck motion to facilitate rolling. Let the purpose of the treatment and the strength of the patient's neck muscles guide you in choosing the correct position. Avoid positions that cause neck pain or general discomfort for the patient.

9.1.3 Therapist Position

To see and control the patient's diagonal neck motion, the best position is on the extension side of center. For example, when the patient moves in the diagonal of right flexion-left extension, stand on the patient's left. For the other diagonal, stand on the patient's right. When the patient is supine or sidelying, stand behind the patient. When the patient is prone, sitting, or standing, you may be either in front or behind the patient. Wherever you stand, align your arms and hands with the diagonal motion.

The body mechanics are essential to guide the head and neck in the right direction. Too little body mechanics, especially in the extension direction, give too little extension movement and maybe too much rotation.

9.1.4 Grips

The grips for neck patterns are on the chin and head. The grip on the chin controls the *short* neck flexion or extension and the rotation. Give the pressure in the center of the chin to avoid side loads on the temporomandibular joint. The grip on the head controls the *long* neck flexion or extension, the rotation, and the lateral motion. The grip is just a little off center on the side of the lateral motion and rotation.

Grip on the patient's chin with the hand that is on the side of extension. Your other hand goes on the patient's head.

Example: The patient is sitting and moving in the diagonal of left flexion-right extension. You are standing behind the patient on the right. Use your right hand on the chin, your left hand on the head.

Example: The patient is prone on the elbows and moving in the diagonal of left flexion-right extension. You are standing in front on the patient's right. Use your left hand on the chin, your right hand on the head. (Your left hand is on the side of the patient's neck extension because you are facing the patient.)

9.1.5 Resistance

Keep the resistance to neck motion within the patient's ability to move or hold without pain or strain. Give resistance to the chin along the line of the mandible. Traction outward to resist flexion, push in to resist extension.

Resist the rotation, lateral motion and the anterior or posterior motion with your proximal hand (on the head).

9.1.6 Normal Timing

The normal timing of the neck patterns is from distal (chin motion) to proximal (neck motion). In the flexion and extension patterns, the head moves through the diagonal in a straight line with rotation occurring throughout the motion. The upper cervical spine moves the chin through its full range of flexion (tucking) or extension (lifting) first. The other joints then move the head through the remaining motion. The rotation occurs smoothly throughout the motion.

9.2 Flexion-Left Lateral Flexion-Left Rotation (Fig. 9.1)

Patient Position. The patient is sitting. You are standing behind the patient to the right of center.

Grip. Put the finger tips of your right hand under the patient's chin. Hold the top of the patient's head with your left hand, just left of center. Your left hand and fingers point in the line of the diagonal. Give the resistance with the fingers and palm of that hand. To apply traction with your proximal hand, hook the carpal ridge under the patient's occiput and lift in the line of the diagonal.

Elongated Position. The chin is elevated and the neck elongated. The extension is evenly distributed among the cervical and upper thoracic vertebrae. The head is rotated and tilted to the right. The chin, nose, and crown of the head are all on the right side of the patient's midline. You should see and feel that the anterior soft tissues on the left side of the patient's neck are taut. None of the vertebral joints

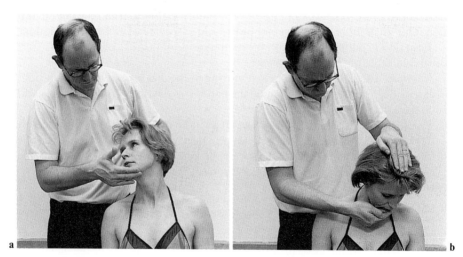

a　　　　　　　　　　　　　　　　　　　　　　　　　　　　　b

Fig. 9.1 a, b. Neck: flexion – left lateral flexion – left rotation in a sitting position

should be in a close-pack position. If you give traction through the neck, the patient's trunk lengthens and rotates to the right.

Body Position and Mechanics. Stand behind the patient, slightly to the right. Your shoulders and pelvis face the diagonal, your arms are aligned with the motion. Allow the patient's motion to pull your weight forward.

Traction. Apply gentle traction by elongating the entire pattern.

Command. "Tuck your chin in. Bend you head down. Look at your left hip."

Movement. The patient's mandible depresses as the chin tucks with rotation toward the left. The neck flexes, following the line of the mandible, bringing the patient's head down towards the chest.

Resistance. Your right hand on the patient's chin gives traction along the line of the mandible and resists the rotation to the left. Your left hand on the patient's head gives a rotational force to the head back toward the starting position. To give traction with this hand, hook the carpal ridge of your hand under the patient's occiput.

End Position. The patient's head, neck, and upper thoracic spine are fully flexed. The rotation and lateral flexion bring the nose, the chin, and the crown of the head to the left of the midline. The patient's nose points towards the left hip.

Alternative Patient Positions. The patient may be prone on the elbows (with the therapist standing behind, Fig. 9.2; or with the therapist standing in front, Fig. 9.3), supine (Fig. 9.4), or in a sidelying position (Fig. 9.5).

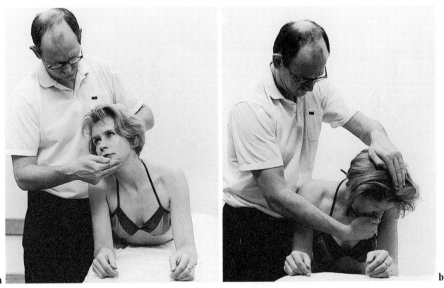

Fig. 9.2 a, b. Neck: flexion – left lateral flexion – left rotation lying prone supported on the elbows

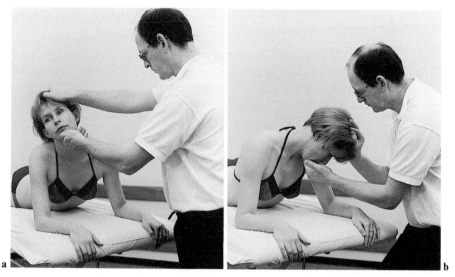

Fig. 9.3 a, b. Neck: flexion – left lateral flexion – left rotation lying prone supported on the elbows with therapist in front

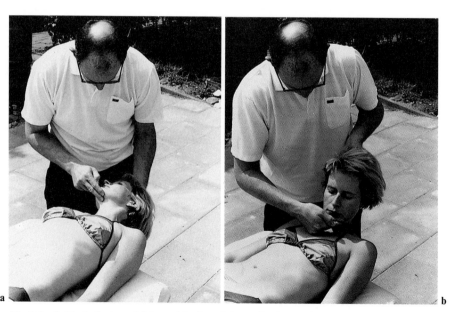

Fig. 9.4 a, b. Neck: flexion – left lateral flexion – left rotation in a supine position

146

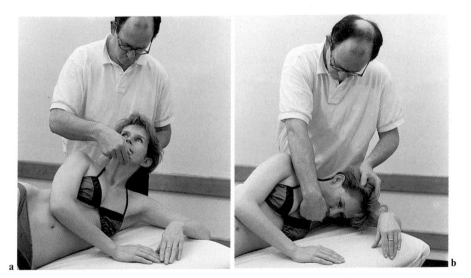

Fig. 9.5 a, b. Neck: flexion – left lateral flexion – left rotation lying on the side

9.3 Extension-Right Lateral Flexion-Right Rotation (Fig. 9.6)

Patient Position. The patient is sitting. You are standing behind the patient to the right of center.

Grip. Put your right thumb on the center of the patient's chin. Hold the top of the patient's head with your left hand, just right of center. Your left hand and fingers point in the line of the diagonal. With this hold, give the resistance with the palm

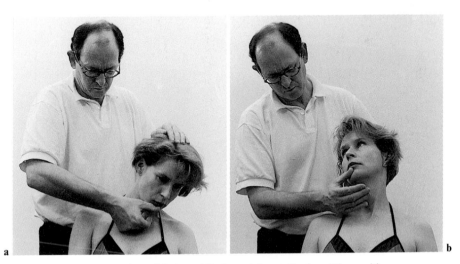

Fig. 9.6 a, b. Neck: extension – right lateral flexion – right rotation in a sitting position

147

and carpal ridge of your hand. To traction with your proximal hand, hook the carpal ridge under the occiput.

Elongated Position. The chin is tucked and the neck flexed. The head is rotated and tilted to the left. The patient's chin, nose, and crown of the head are all on the left side of the midline. You should see and feel that the posterior soft tissues on the right side of the patient's neck are taut. None of the vertebral joints should be in a close-pack position. If you give traction through the neck, the patient's trunk flexes and rotates to the left.

Body Position and Mechanics. Stand behind the patient, slightly to the right. Your shoulders and pelvis face the diagonal, your arms are aligned with the motion. Allow the patient's motion to push your weight back, and the therapist moves away from the patient. Doing this, you get the right extension and lateral-flexion movement.

Traction. Apply gentle traction to the skull to elongate the neck. Gently compress on the chin through the line of the mandible.

Command. "Lift your chin. Lift your head. Look up."

Movement. The patient's mandible protrudes and the chin lifts with rotation toward the right. The neck and upper thoracic spine extend, following the line of the mandible. The patient's neck and upper spine elongate as the head comes up.

Resistance. Your right hand on the patient's chin compresses along the line of the mandible and resists rotation to the right. Your left hand on the patient's head gives a rotational force to the head back toward the starting position. Use traction through the head during the first part of the motion. As the neck approaches the extended position, you may apply gentle compression through the top of the patient's head.

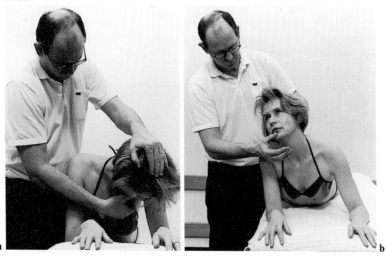

Fig. 9.7 a, b. Neck: extension – right lateral flexion – right rotation lying prone supported on the elbows

148

End Position. The patient's head, neck, and upper thoracic spine are extended with elongation. The rotation and lateral flexion bring the nose, the chin, and the crown of the head to the right of the midline. *Caution:* Do not allow excessive extension in the mid-cervical area. The neck must elongate, not shorten.

Alternative Patient Positions. The patient may be prone on the elbows (with the therapist standing behind, Fig. 9.7; or with the therapist standing in front), supine, or in a sidelying position.

9.4 Neck Rotation to the Right

Patient Position. The patient is sitting. Stand at the patient's left side (opposite the direction of rotation).

Grip. Place your right hand on the right side of the patient's head (by the right ear). Put the fingers of your left hand under the patient's chin.

Starting Position. The patient's head is in the midline.

Body Position and Mechanics. Stand at the patient's left side and face toward the direction of rotation.

Command. "Tuck your chin and turn your head to the right." "Try to put your chin behind your right shoulder." "Look over your right shoulder."

Movement. The mandible retracts as the chin tucks in. The neck and upper trunk extend with rotation and lateral flexion to the right.

Resistance. Your left hand on the chin resists the short neck flexion, the lateral flexion, and the rotation. Your right hand, by the right ear, resists the long neck extension and rotation.

Normal Timing. The chin tucks first. The other motions occur simultaneously.

Alternative Patient Positions. The patient may be prone, prone on the elbows, or supine.

9.5 Neck for Trunk

When the neck is strong and pain-free you can use it as a handle to exercise the trunk muscles. Both static and dynamic techniques work well. If there is a chance that motion will cause pain, pre-position the neck in the desired end range and use static contractions. *Note:* The head and neck are the handle, the action happens in the trunk.

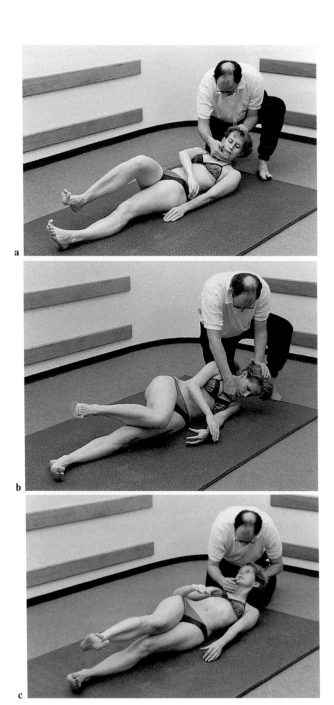

Fig. 9.8 a–c. Neck for trunk flexion and extension. ***a, b*** Neck flexion for rolling forward. ***c*** Neck extension for rolling backward

When using neck flexion patterns, the main component of resistance is traction. With extension patterns, gentle compression through the crown of the head will facilitate trunk elongation with the extension.

9.5.1 Neck for Trunk Flexion and Extension

With the patient supine, use the neck to facilitate rolling forward. (Fig. 9.8a, b). If the patient has good potential trunk strength use the neck to facilitate a supine-to-sitting motion. With the patient sidelying or prone use neck extension to facilitate rolling back (Fig. 9.8c). Resist static neck flexion and extension patterns with the patient's head in the midline to facilitate static contractions of the trunk muscles in sitting. To challenge the patient's sitting balance use reversal techniques, either static or with small reversing motions.

When you exercise the patient in standing give gentle resistance to the neck patterns. Combine this with resistance at the shoulder or pelvis.

9.5.2 Neck for Trunk Right Lateral Flexion

This activity can be used in all positions. The shortening on the working side results in concurrent lengthening on the other side. The trunk lateral flexion is facilitated by the chin tuck (short flexion), rotation, and lateral flexion. After the head is positioned, all further motion occurs in the trunk. This pattern can be done with a flexion or an extension bias.

9.5.2.1 Right Lateral Flexion with Flexion Bias (Fig. 9.9a–c)

Begin with the patient's chin tucked and the head turned so the chin aims towards the front of the right shoulder.

Body Position and Mechanics. Stand at the patient's left side, opposite to the direction of rotation.

Grip. Place your right hand on the right side of the patient's head (by the right ear). Put left hand under the patient's chin.

Alternative Body Position and Grip. Stand so the patient turns toward you. Put your left hand by the patient's right ear. Place the fingers of your right hand under the patient's chin.

Command. Preparation: "Turn your head to the right and put your chin here (touching the front of the right shoulder)."

Command: "Keep your chin on your shoulder, don't let me move your head." "Now pull your chin farther to your shoulder" "Now do it again."

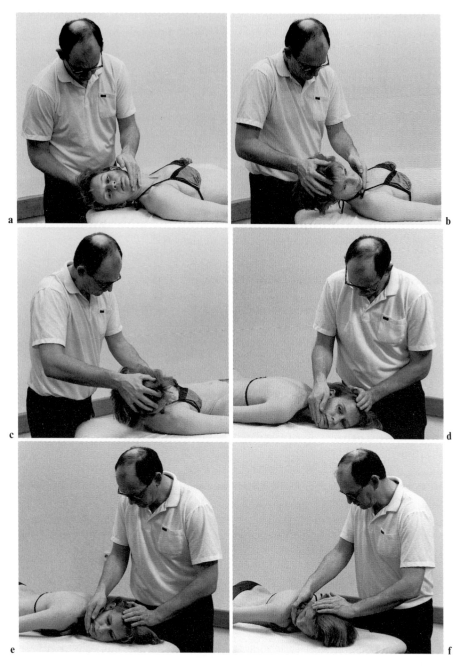

Fig. 9.9 a–f. Neck for trunk right lateral flexion. *a–c* In a supine position with flexion bias. *d–f* In a prone position with extension bias

Resistance. Your distal hand (chin) resists short neck flexion, rotation and lateral flexion. Your proximal hand resists long neck extension, rotation and lateral flexion.

Motion. The upper trunk side-bends to the right, the right shoulder moves towards the right ilium. The motion includes flexion and right rotation.

9.5.2.2 Right Lateral Flexion with Extension Bias

Begin with the patient's chin tucked and the head turned so the chin aims towards the back of the right shoulder (Fig. 9.9 d, e).

Body Position and Mechanics. Stand at the patient's left side opposite to the direction of rotation.

Grip. The grip is the same as above.

Alternative Body Position and Grip. These are the same as above.

Command. Preparation: "Turn your head to the right and try to put your chin behind your right shoulder."

Command. "Keep your chin on your shoulder and your ear back; don't let me move your head." "Now pull your chin farther behind your shoulder." "Now do it again."

Resistance. Your distal hand (chin) resists short neck flexion, rotation, and lateral flexion. Your proximal hand resists long neck extension, rotation, and lateral flexion.

Note: When your patient is prone, the rotational resistance is away from you (toward the front of the patient).

Motion. The upper trunk side-bends to the right with extension, the right shoulder moves towards the back of the right ilium. The motion includes trunk extension and right rotation.

Reference

Kendall FP, McCreary EK (1983) Muscles, testing and function. Williams and Wilkins, Baltimore

10 The Trunk

10.1 Introduction

A strong trunk is essential for good function. The trunk is the base that supports extremity motions. For example, supporting trunk muscles contract synergistically with arm motions (Angel and Eppler 1967). With the trunk able to move and stabilize effectively, patients gain improved control of their arms and legs.

Strengthening the muscles of the trunk is only one reason for using the trunk patterns in patient treatment. Some other uses for these patterns are:

- Resisting the lower trunk patterns provides irradiation for indirect treatment of the neck and scapular muscles.
- Continuing the upper trunk patterns exercises the patient's hips by moving the pelvis on the femur.
- Resist trunk activity to produce irradiation into the other extremities. For example, when you resist the lower extremities for trunk flexion and extension, the patient's arm muscles work to help with stabilization.

Use of the scapula and pelvis to facilitate activity of the trunk muscles is covered in Chap. 6. Chapter 9 describes using the neck to facilitate trunk motion. In this chapter we focus on using the extremities to exercise the trunk muscles.

10.1.1 Diagonal Motion

The trunk flexion and extension patterns have the same three motion components as the other patterns: flexion or extension, lateral flexion, and rotation. The axis of motion for the flexion and extension patterns runs approximately from the coracoid process to the opposite anterior superior iliac spine (ASIS). The lateral flexion (side-bending) patterns have three components as well. The emphasis in this activity is lateral trunk bending with rotation and flexion or extension.

In this chapter we illustrate and describe the diagonal of flexion to the left, extension to the right. To work with the other diagonal, reverse the words "left" and "right" in the instructions.

Trunk Flexion to the Left – Left Lateral Flexion – Left Rotation

Movement	Muscles: principal components (Kendall and McCreary 1983)
Chopping to the left	Left external oblique, rectus abdominis, right internal oblique
Bilateral lower extremity flexion to the left	Left internal oblique, rectus abdominis, right external oblique muscle

Trunk Extension to the Right – Right Lateral Flexion – Right Rotation

Lifting to the right	All the neck and back extensor muscles, left multifidi and rotatores
Bilateral lower extremity extension to the right	All the back and neck extensor muscles, right quadratus lumborum, left multifidi and rotatores

Trunk Lateral Flexion to the Right

With extension bias	Quadratus lumborum, iliocostalis lumborum, longissimus thoracis, latissimus dorsi (when arm is fixed)
With flexion bias	Right internal oblique, right external oblique.

10.1.2 Patient Position

The patient can be in any position when exercising the trunk muscles. We have found that the following combinations give good results:

– Supine: upper and lower trunk flexion and extension, lateral flexion
– Sidelying: upper and lower trunk flexion and extension
– Prone: upper trunk extension
– Sitting: upper trunk flexion and extension, upper trunk side bending using the neck, irradiation from the upper trunk into lower trunk and hip motions

We introduce the patterns with the patient supine and show variations of position later in the chapter.

10.1.3 Resistance

Block the initial motion of the extremities until you feel or see the patient's trunk muscles contract. Then allow the extremities to move, maintaining enough resistance to keep the trunk muscles contracting.

10.1.4 Normal Timing

With these combination patterns, the extremities start the motion while the trunk muscles stabilize. After the extremities have moved through range, the trunk completes its motion.

10.1.5 Timing for Emphasis

Lock in the extremities at the end of their range of motion. Use them as a handle to exercise the trunk motion.

10.2 Chopping and Lifting

These combination patterns use bilateral, asymmetrical, upper extremity patterns combined with neck patterns to exercise *trunk muscles.* The arms are resisted as a unit. Successful use of these combinations requires that at least one of the arms must be strong.

Note: You may use any elbow motion with the shoulder patterns. We have used the straight arm patterns in these illustrations.

10.2.1 Chopping

Bilateral asymmetrical upper extremity extension with neck flexion can be used *for* trunk flexion, as shown here. Other uses for the chopping pattern are:

– Facilitating functional motions such as rolling forward or coming to sitting
– Exercising hip flexion when the muscles of trunk flexion are strong

Chopping to the left is illustrated in Figs. 10.1, 10.2. Its components are:

1. Left arm (the lead arm): extension-abduction-internal rotation.
2. Right arm (the following arm): extension-adduction-internal rotation. The following (right) hand grips the lead (left) wrist.
3. Neck: flexion to the left.

Patient Position. The patient is supine and close to the left side of the table.

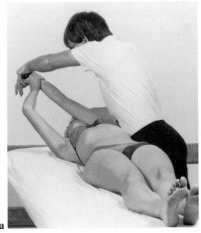 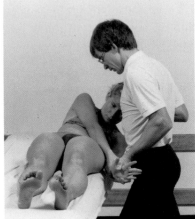

a   b

Fig. 10.1 a, b. Chopping to the left in a supine position

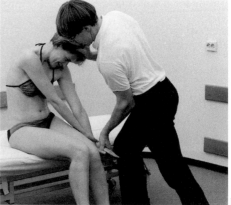

a   b

Fig. 10.2 a, b. Chopping to the left in a sitting position

Body Position and Mechanics. Stand in a stride position on the left side of the table facing toward the patient's hands. This is the same position used to resist the single arm pattern of extension-abduction-internal rotation. Let the patient's motion push your weight back. As the patient's arm nears the end of the range, turn your body so you face the patient's feet.

Grip. *Distal hand:* Your left hand grips the patient's left hand (leading hand). Use the normal distal grip for the pattern of extension-abduction-internal rotation.

Proximal hand: Place your right hand on the patient's forehead with your fingers pointing towards the crown.

Elongated Position. The patient's left arm is in flexion-adduction-external rotation. The right hand grips the left wrist with the right arm in modified flexion-abduction-external rotation. The patient looks at the left hand, putting the neck in modified extension to the right (see Fig. 10.1 a).

Stretch. Traction the left arm and scapula until you feel the trunk muscles elongate. Continue the traction to give stretch to the arms and the trunk.

Command. "Push your arms down to me and lift your head. Now keep your arms down here and push some more." "Reach for your left knee."

Movement. The patient's left arm moves through the pattern of extension-abduction-internal rotation with the right arm following into extension-adduction-internal rotation. The patient's head and neck come into flexion to the left. At the same time, the patient's upper trunk begins to move into flexion with rotation and lateral flexion to the left.

Resistance. The major resistance is to the arm motion and through the arms into the trunk. The resistance to the head is light and serves mainly to guide the head and neck motion.

158

Use resistance to hold back on the beginning arm motion until you feel and see the abdominal muscles begin to contract. Then allow the arms and head to complete their motion against enough resistance to keep the trunk flexor muscles contracting.

End Position. The left arm is extended by the patient's side and the patient's neck is in flexion to the left. The upper trunk is flexed to the left as far as the patient can go.

Normal Timing. The abdominal muscles begin to contract as soon as the arms and head begin their motion. By the time the arms and head have finished their movement, the upper trunk if flexed with left rotation and left lateral flexion.

Timing for Emphasis. Lock in the arms at their end range using approximation and rotational resistance. Using the stable arms as a handle, exercise the trunk flexion. *Note:* In the end position the arms and head are the handle and do not move. Only the trunk moves as you exercise it.

When working on the mats use chopping to help the patient roll forward or go from supine to sitting. The rotational resistance and approximation through the arms promote and resist the patient's motion to sitting. Shift the angle of resistance slightly to the side to get the patient to roll.

Alternative Position. Sitting: your goal can be flexion of the trunk with gravity assistance or flexion of the hips with irradiation from the arms and trunk. Use this position to train the trunk and hip flexor muscles in eccentric work.

10.2.2 Lifting

Bilateral asymmetrical upper extremity flexion with neck extension can be used *for* trunk extension, as shown here. Other uses for the lifting pattern are:

– Exercising hip extension when the trunk extensor muscles are strong
– Facilitating functional motions such as rolling backward or coming to erect sitting

Lifting to the left is illustrated in Fig. 10.3. Its components are:

1. Left arm (the lead arm): flexion-abduction-external rotation.
2. Right arm (the following arm): flexion-adduction-external rotation. The following (right) hand grips the lead (left) wrist.
3. Neck: extension to the left.

Patient Position. The patient is supine and close to the left side of the table (Fig. 10.3 a, b).

Body Position and Mechanics. Stand in a stride position on the left at the head of the table facing towards the patient's hands. Let the patient's motion push your weight back. As the patient's arm nears the end of the range, step back in the line of the diagonal.

Grip. *Distal hand:* Your left hand grips the patient's left hand (leading hand). Use the normal distal grip for the pattern of flexion-abduction-external rotation.

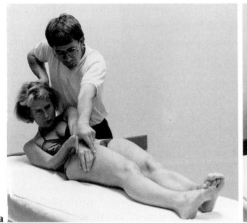
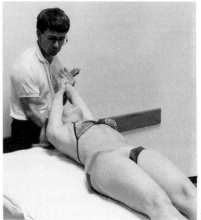

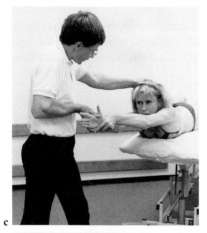

Fig. 10.3 a–e. Lifting. **a, b** Lifting to the left in a supine position. **c** Lifting to the right in a prone position. **d, e** Lifting to the left in a sitting position

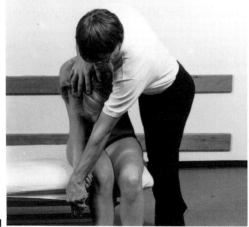

160

Proximal hand: Place your right hand on the crown of the patient's head with your fingers pointing towards the left side of the patient's neck.

Elongated Position. The patient's left arm is in extension-adduction-internal rotation. The right hand grips the left wrist with the right arm in modified extension-abduction-internal rotation. The patient looks at the left hand putting the neck in flexion to the right (Fig. 10.3 a).

Stretch. Traction the left arm and scapula until you feel the arm and trunk muscles elongate. Continue the traction to give stretch to the arms and the trunk. Traction the head to elongate the neck extensor muscles.

Command. "Lift your arms up to me and push your head back. Follow your hands with your eyes. Now keep your arms and head back here and push some more."

Movement. The patient's left arm moves through the pattern of flexion-abduction-external rotation with the right arm following into flexion-adduction-external rotation. The patient's head and neck come into extension to the left. At the same time the patient's upper trunk begins to move into extension with rotation and lateral flexion to the left.

Resistance. The resistance is to the arm and head motion and through them into the trunk. Use resistance to hold back on the beginning arm and head motion until you feel and see the back extensor muscles begin to contract. Then allow the arms and head to complete their motion against enough resistance to keep the trunk extensor muscles contracting.

End Position. The arms are fully flexed with the left arm by the patient's left ear. The patient's head is extended to the left. The trunk is extended and elongated to the left. The extension continues down to the legs if the patient's strength permits.

Normal Timing. The back extensor muscles begin to contract as soon as the arms and head begin their motion. By the time the arms and head have finished their movement, the trunk is extended with left rotation and left lateral flexion.

Timing for Emphasis. Lock in the arms and head at their end range. Lock in the arms using resistance to rotation and approximation, the neck with resistance to rotation and extension. Use the arms and head as a handle to exercise the trunk extension (elongation). Neither the arms nor the head should move while the trunk is exercising. Use lifting when the patient is lying on the mats with the goal of rolling backward.

Alternative Positions. *Prone:* exercise in the end range against gravity. This position is particularly good with stronger and heavier patients (Fig. 10.3 c).
   Sitting: your goal is elongation of the trunk. Do not allow the patient to move into hyperlordosis in the cervical or lumbar spine.

Use lifting to facilitate moving from a bent (flexed) to an upright (extended) position. Lifting is also good for teaching the patient erect posture (Fig. 10.3 d, e).

10.3 Bilateral Leg Patterns for the Trunk

These combinations use bilateral, asymmetrical, lower extremity patterns to exercise *trunk muscles.* Hold the legs together and resist them as a unit. Successful use of these combinations requires that at least one of the legs be strong.

Note: You may use any knee motion with the hip patterns. The usual combination is hip flexion with knee flexion and hip extension with knee extension.

10.3.1 Bilateral Lower Extremity Flexion, with Knee Flexion, for Lower Trunk Flexion (Right) (Fig. 10.4 a, b)

Position at Start. Position the patient close to the edge of the table. The patient's legs are together with the left leg in extension-abduction-internal rotation and the right leg in extension-adduction-external rotation.

Body Mechanics. Stand in a stride facing the diagonal. Lean back to elongate and stretch the pattern. As the patient's legs move up into flexion, step forward with your rear leg. Use your body weight to resist the motion.

Grip. *Distal hand:* Your left hand holds both of the patient's feet with contact on the dorsal and lateral surfaces of both feet. Do not put your finger between the patient's feet. If the feet are too large for your grasp, cross one foot partially over the other to decrease the width.
   Proximal hand: Your right arm is underneath the patient's thighs. Hold the thighs together with this arm.

Elongated Position. The trunk is extended and elongated to the left with left rotation and side-bending. *Caution:* Avoid hyperextension in the lumbar spine.

Stretch. Traction and rotate the legs to elongate and stretch the lower extremity and trunk flexor muscles.

Command. "Feet up, bend your legs up and away. Bring your knees to your right shoulder."

Movement. As the feet dorsiflex the trunk flexor muscles begin to contract. The legs flex together: the right leg into flexion-abduction-internal rotation, the left leg into flexion-adduction-external rotation. When the legs reach the end of their range, the motion continues as lower trunk flexion with rotation and side-bending to the right.

Resistance. *Distal hand:* Resist trunk and hip rotation with pressure on the lateral border of the right foot. Resist the knee motion with this hand as you did with the single leg patterns. If the knees remain straight, give traction through the line of the tibia.

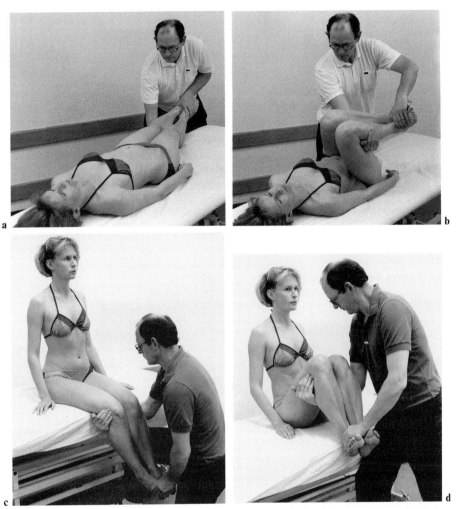

Fig. 10.4 a–d. Bilateral lower extremity flexion with knee flexion for lower trunk flexion:
a, b supine, *c, d* sitting

Proximal hand: Continue to hold the thighs together with this arm. Use your hand to resist rotation and lateral motion with pressure on the lateral border of the thigh. Give traction through the line of the femur. *Caution:* Too much resistance to hip flexion will cause the lumbar spine to hyperextend.

End Position. The right leg is in full flexion-abduction-internal rotation, the left leg in full flexion-adduction-external rotation. The lower trunk is flexed with rotation and lateral flexion to the right.

Normal Timing. As the legs begin to flex the trunk flexor muscles contract. After the hips have reached their end range, the motion continues with lower trunk flex-

163

ion. *Caution:* Do not allow the lumbar spine to be pulled into hyperextension. Start with the legs flexed if the trunk flexor muscles cannot stabilize the pelvis at the beginning of the motion.

Timing for Emphasis. Lock in the lower extremities in their end position. Use the legs as a handle to exercise the trunk motion. You may use static or dynamic exercises. *Note:* In the end position the legs are the handle. Only the pelvis moves while you exercise the trunk.

To exercise neck and upper trunk flexion, use prolonged static contraction of the legs and lower trunk muscles. Using the legs and lower trunk in this way works well when the patient's arms are too weak to use for upper trunk exercise. This combination is also useful when the patient has pain in the neck or upper trunk.

Alternative Positions. Use this lower extremity combination on the mats to facilitate rolling from supine to side-lying or in a sitting position (Fig. 10.4c, d).

10.3.2 Bilateral Lower Extremity Extension, with Knee Extension, for Lower Trunk Extension (Left) (Fig. 10.5)

Position at Start. Position the patient close to the left side of the table.

Body Mechanics. Stand in a stride position facing the diagonal. Lean forward to stretch the pattern. As the patient's legs move into extension, step back with your forward leg. Use your body weight to resist the motion.

Grip. *Distal hand:* Your left hand holds both of the patient's feet with contact on the plantar and lateral surfaces close to the toes. If the feet are too large for your grasp, cross one foot partially over the other to decrease the width.

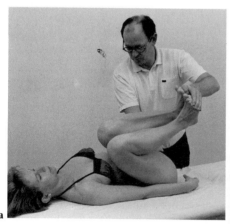
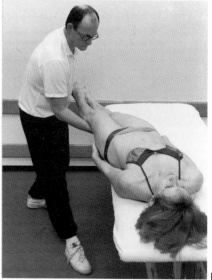

Fig. 10.5 a, b. Bilateral lower extremity extension with knee extension for lower trunk extension

Proximal hand: Your right arm is underneath the patient's thighs. Hold the thighs together with this arm.

Elongated Position. The patient's legs are flexed to the right. The right leg is in flexion-abduction-internal rotation with knee flexion, the left leg in flexion-adduction-external rotation with knee flexion. The lower trunk is flexed with rotation and lateral flexion to the right.

Stretch. Use traction with rotation through the thighs to increase the trunk flexion to the right.

Command. "Toes down, kick down to me."

Movement. As the feet plantar flex the trunk extensor muscles begin to contract. The legs extend together, the left leg into extension-abduction-internal rotation, right leg into extension-adduction-external rotation. When the legs reach the end of their range, the motion continues as lower trunk elongation with rotation and side-bending to the left.

Resistance. *Distal hand:* Resist trunk and hip rotation with pressure on the lateral border of the left foot. Resist the knee extension with this hand as you did with the single leg patterns. If the knees remain straight, give approximation through the line of the tibia.
   Proximal hand: Continue to hold the thighs together with this arm as you resist the hip motions.

End Position. The left leg is in full extension-abduction-internal rotation, the right leg in full extension-adduction-external rotation. The lower trunk is elongated with rotation and lateral flexion to the left.

Normal Timing. The trunk extensor muscles contract as soon as the legs begin their motion. By the time the leg motion is completed the trunk is in full elongation. *Caution:* The end position is trunk elongation, not lumbar spine hyperextension.

Timing for Emphasis. To exercise the neck and upper trunk extension, use prolonged static contraction of the legs and lower trunk muscles. Using the legs and lower trunk in this way works well when the patient's arms are too weak to use for upper trunk exercise. This combination is also useful when the patient has pain in the neck or upper trunk.

Alternative Positions. Use this lower extremity combination on the mats to facilitate rolling from sidelying or prone to supine.

10.3.3 Trunk Lateral Flexion

The lateral flexion pattern can be done with a trunk flexion bias or an extension bias. To exercise the motion, use the bilateral leg flexion or extension patterns with full hip rotation.

10.3.3.1 Left Lateral Flexion with Flexion Bias

Begin with the legs in the shortened range of bilateral lower extremity flexion to the left.

Command. "Swing your feet away from me (to the left)." If you are working with straight leg patterns, a good command is: "Turn your heels away from me."

Resistance. With your proximal hand give traction through the thighs to lock in the hip flexion. Lateral pressure resists the lateral hip motion. With your distal hand lock in the knees and feet and resist the hip rotation.

Movement. The hips and knees are flexed to the left. As the hips rotate left past the groove of the flexion pattern the lumbar spine side-bends to the left and the pelvis moves up towards the ribs.

10.3.3.2 Right Lateral Flexion with Extension Bias

We can exercise this motion in the lengthened or the shortened range of the leg patterns.

In the Lengthened Range
Begin with the patient's legs in full flexion to the left (the lengthened range of bilateral lower extremity extension to the right).

Body Mechanics. Stand in a stride position by the patient's left shoulder. Use your body weight to resist the leg and trunk motion.

Command. "Swing your feet to the right and push your legs away." If you wish, ask for a static rather than a dynamic contraction of the hip and knee extension.

Resistance. With your proximal hand resist the hip extension and lateral motion. Your distal hand locks in the knee and foot motion and resists the dynamic hip rotation.

Motion. The hips rotate fully to the right. The lumbar spine extends and side-bends right. *Note:* You may allow a few degrees of hip and knee extension to the right.

In the Shortened Range
Body Mechanics. Stand on the right and use your body as you did for the pattern of trunk extension to the right.

Command. "Kick to me. Turn your heels toward me."

Resistance. Give the same resistance as you did for trunk extension. Allow full hip rotation.

Motion. The patient's legs extend to the right with full hip rotation. The lumbar spine extends and side-bends right.

10.4 Combining Patterns for the Trunk

You can combine the upper and lower trunk patterns to suit the needs of the patient. When treating an adult patient work in positions where you can handle the patterns comfortably. You may pre-position the patient's arms and legs in the end range of the patterns you are exercising. Choose techniques that are suited to the patient's needs and strengths. Some trunk combinations are:

1. Upper and lower trunk flexion:
   a) With counterrotation of the trunk
      Chopping to the left with bilateral leg flexion to the right (Fig. 10.6).
   b) Without trunk counterrotation
      Chopping to the left with bilateral leg flexion to the left.

2. Upper and lower trunk extension:
   a) With counterrotation of the trunk
      Lifting to the right with bilateral leg extension to the left. Use a static contraction of the lower extremity extension pattern from the flexed position (Fig. 10.7).
   b) Without counterrotation of the trunk
      Lifting to the left with bilateral leg extension to the left.

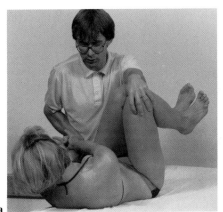
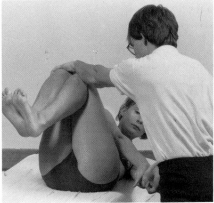

a

b

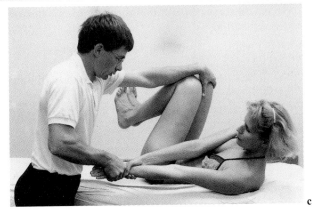

Fig. 10.6 a–c. Trunk combination: chopping to the left and bilateral leg flexion to the right

c

167

Fig. 10.7. Trunk combination: upper and lower trunk extension by lifting to the right and bilateral leg extension to the left

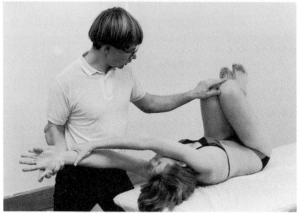

Fig. 10.8. Trunk combination: lifting to the left and bilateral leg flexion to the left

3. Upper trunk flexion with lower trunk extension:
   a) With trunk counterrotation
      Chopping to the left with bilateral leg extension to the right.
   b) Without trunk counterrotation
      Chopping to the left with bilateral leg extension to the left.

   Note: Use static contractions of the lower extremity extension pattern from the flexed position.

4. Upper trunk extension with lower trunk flexion:
   a) With trunk counterrotation
      Lifting to the left with bilateral leg flexion to the right.
   b) Without trunk counterrotation
      Lifting to the left with bilateral leg flexion to the left (Fig. 10.8).

References

Angel RW, Eppler WG Jr (1967) Synergy of contralateral muscles in normal subjects and patients with neurologic disease. Arch Phys Med 48: 233–239

Kendall FP, McCreary EK (1983) Muscles, testing and function. Williams and Wilkins, Baltimore

11 Mat Activities

11.1 Introduction: Why do Mat Activities?

The mat program involves the patient in activities incorporating both movement and stability. They range from single movements, such as unilateral scapula motions, to complex combinations requiring both stabilization and motion, such as crawling or knee walking. The activities are done in different positions, for function and to vary the effects of reflexes or gravity. The therapist also chooses positions that can help control abnormal or undesired movements.

When working with an infant it may be necessary to progress treatment using activities that suit the developmental level of the individual. With the adult patient more mature or advanced activities can be used before the more basic activities. The therapist must keep in mind the changing ways in which we accomplish physical tasks as we age (Van Sant 1991).

Functional goals direct the choice of mat activities. An activity, such as getting from supine to sitting, is broken down and the parts practiced. As there are many different ways in which a person can accomplish any activity, treatments should include a variety of movements. For example, to increase trunk and leg strength, the patient may begin treatment with resisted exercises in sitting and side-sitting. The treatment then progress to positions involving more extremity weight bearing. As the patient's abilities increase, exercises that combine balance and motion in bridging, quadruped, and kneeling positions are used. With all functional activities, the patient learns to:

1. Move into a position
2. Stabilize (balance) in that position
3. Combine functional motion with stability of position

Once patients achieve a reasonable degree of competence in an activity they can safely practice on the mats alone or with minimal supervision. Learning and practicing the skills necessary for self-care and gait is easier for the patient when he or she feels secure and comfortable.

11.2 Carrying Out Mat Activities

The therapist should employ all the basic procedures to heighten the patient's capacity to work effectively and with minimum fatigue. *Approximation* promotes stabilization and balance. *Traction* and *stretch* increase the patient's ability to move. Use of correct *grips* and proper *body position* enables the therapist to guide the patient's motion. *Resistance* enhances and reinforces the learning of an activity. Properly graded resistance strengthens the weaker motions. Resisting strong motions provides *irradiation* into the weaker motions or muscles. *Timing for emphasis* enables the therapist to use strong motions to exercise the weaker ones. The *patterns* are used to improve performance of functional activities. All the techniques are suitable for use with mat activities.

11.3 Examples of Mat Activities

These examples of mat activities and exercises are not an all-inclusive list but are samples only. As you work with your patients you will find many other positions and actions to help them achieve their functional goals.

11.3.1 Rolling

Rolling is both a functional activity and an exercise for the entire body. The therapist can learn a great deal about patients by watching them roll. Some people roll using flexion movements, others use extension, and others push with an arm or a leg. Some find it more difficult to roll in one direction than in the other, or from one starting position. The ideal is for individuals to adjust to any condition placed upon them and still be able to roll easily.

The goal of rolling can be strengthening of trunk muscles, increasing the patient's ability to roll, or both. The therapist uses whatever combination of scapula, pelvis, or extremity motions best facilitates and reinforces the desired motions.

11.3.1.1 Scapula

Resistance to either of the anterior scapular patterns facilitates forward rolling. Resisting the posterior scapular patterns facilitates rolling back. Use the appropriate grips for the chosen scapular pattern. To get increased facilitation, patients move their head in the same direction as the scapula.

The command given can be an explicit direction or a simple action command. An explicit direction for rolling using scapular anterior depression would be "pull your shoulder down towards your opposite hip, lift your head, and roll forward." A simple action command for the same motion is "pull down." A simple command for rolling back using posterior elevation is "push back" or "shrug." Telling the patient to look in the direction of the scapula motion is a good command for the head movement.

To start, place the scapula in the elongated range to stretch the scapular muscles. To stretch the trunk muscles, continue moving the scapula farther in the same diagonal until the trunk muscles are elongated. Resist the initial contraction at the scapula enough to *hold back* on the scapular motion until you feel or see the patient's trunk muscles contract. When the trunk muscles begin to contract, allow both the scapula and trunk to move. You can *lock in* the scapula at the end of its

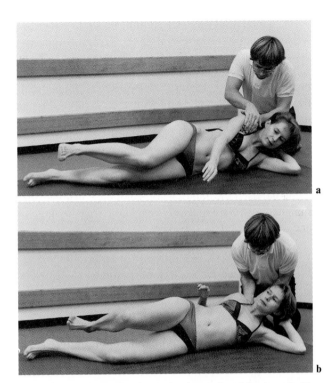

Fig. 11.1 a, b. Using the scapula for rolling: ***a*** forward with anterior depression, ***b*** backward with posterior elevation

range of motion and exercise the trunk muscles and the rolling motion with repeated contractions.

The patterns are:

1 a. Anterior elevation: Roll forward with trunk rotation and extension. Facilitate with neck extension and rotation in the direction of the rolling motion.
1 b. Posterior depression: Roll back with trunk extension, lateral flexion, and rotation. Facilitate with neck lateral flexion and full rotation in the direction of the rolling motion.
2 a. Anterior depression: Roll forward with trunk flexion. Facilitate with neck flexion in the direction of the rolling motion (Fig. 11.1 a).
2 b. Posterior elevation: Roll back with trunk extension. Facilitate with neck extension in the direction of the rolling motion (Fig. 11.1 b).

11.3.1.2 Pelvis

Resistance to pelvic anterior elevation facilitates rolling forward, resisting posterior depression facilitates rolling back. Use the appropriate grips for the chosen pattern. Ask for neck flexion to reinforce rolling forward, extension for rolling back.

171

The commands for pelvic motion are similar to those for the scapula. For rolling forward using anterior elevation the explicit command would be "pull your pelvis up and roll forward". The simple command for the same motion is "pull." For rolling back using posterior depression a specific command would be "sit down into my hand and roll back." The simple command for that action is "push." Telling the patient to look in the direction of the desired rolling motion is a good command for the head movement.

To start, place the pelvis in its elongated range. To stretch the trunk further, continue moving the pelvis in the same diagonal until the trunk is completely elongated. Resist the initial contraction at the pelvis until you feel or see all of the desired trunk muscles contract. Then allow both the pelvis and trunk to move. You can *lock in* the pelvis at the end of its range of motion and exercise the rolling motion with repeated contractions.

The patterns are:

1a. Anterior elevation: Roll forward with trunk flexion (Fig. 11.2a).
1b. Posterior depression: Roll back with trunk extension (Fig. 11.2b).

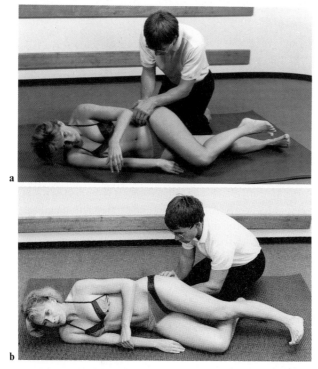

Fig. 11.2a, b. Using the pelvis for rolling: *a* forward with anterior elevation, *b* backward with posterior depression

11.3.1.3 Scapula and Pelvis

A combination for rolling forward is: pelvis in anterior elevation, scapula in anterior depression (Fig. 11.3).

A combination for rolling backward is: pelvis in posterior depression, scapula in posterior elevation (Fig. 11.4).

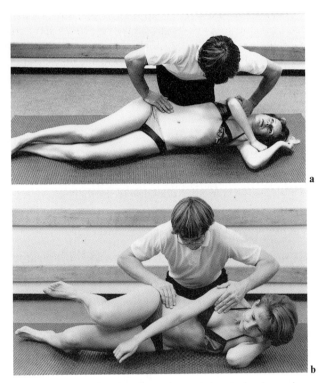

Fig. 11.3 a, b. Rolling forward with pelvis in anterior elevation and scapula in anterior depression

Fig. 11.4 a

b

Fig. 11.4 a, b. Rolling backward with pelvis in posterior depression and scapula in posterior elevation

11.3.1.4 Upper Extremities

Use the arm to strengthen the trunk muscles and to facilitate rolling in the same way as for the scapula. Adduction (anterior) patterns facilitate rolling forward, abduction (posterior) patterns facilitate rolling back. The patient's head should move with the arm.

Your distal grip is on the hand or distal forearm and can control the entire extremity. Your proximal grip can vary: a grip on or near the scapula is often the most effective. Your proximal hand can also be used to guide and resist the patient's head motion.

The commands you use can be specific or simple. For rolling forward using the pattern of extension-adduction the specific command may be "squeeze my hand and pull your arm down to your opposite hip. Lift your head, and roll." The simple command would be "squeeze and pull, lift your head." For rolling back using the pattern of flexion-abduction the specific command might be "wrist back, lift your arm up and follow your hand with your eyes. Roll back." The simple command would be "lift your arm up and look at your hand."

Take the patient's arm into the elongated range and traction to stretch the arm and scapular muscles. Further elongation with traction will elongate or stretch the synergistic trunk muscles. Hold back on the inital arm motion until you feel or see the patient's trunk muscles contract, then allow the arm and trunk to move. You can lock in the patient's arm at any strong point in its range of motion, then exercise the trunk muscles and the rolling motion with repeated contractions. Approximation through the arm with resistance to rotation works well to lock in the arm towards the end of its range.

The patterns are:

One arm

1 a. Flexion-adduction (Fig. 11.5 a)

1 b. Extension-abduction (Fig. 11.5 b): Rolling back with trunk extension, lateral flexion, and rotation. Facilitate with neck lateral flexion and full rotation in the direction of the rolling motion.

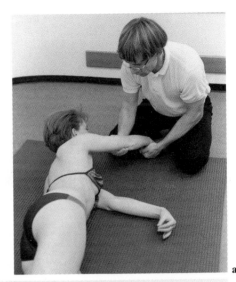

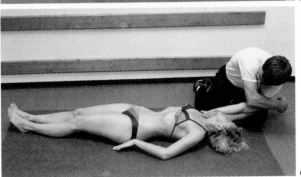

Fig. 11.5 a–c

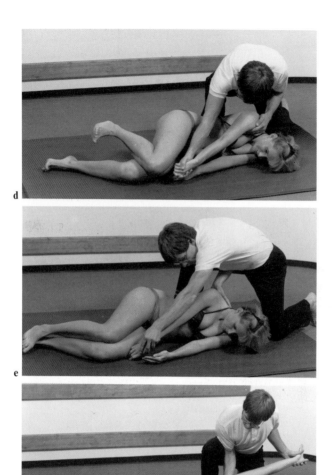

Fig. 11.5 a–f. Using one arm for rolling: *a* forward with flexion – adduction, *b* backward with extension – abduction, *c, d* forward with extension – adduction, *e, f* backward with flexion – abduction

2 a. Extension-adduction (Fig. 11.5 c, d): Rolling forward with trunk flexion. Facilitate with neck flexion in the direction of the rolling motion.
2 b. Flexion-abduction (Fig. 11.5 e, f): Rolling back with trunk extension. Facilitate with neck extension in the direction of the rolling motion.

Both arms
3 a. Chopping (Fig. 11.6 a): Rolling forward with trunk flexion.
3 b. Lifting (Fig. 11.6 b, c): Rolling back with trunk extension.

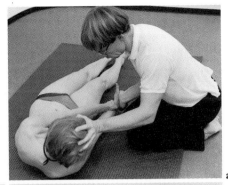

Fig. 11.6 a–c. Using both arms for rolling: *a* forward with chopping, *b, c* backward with lifting

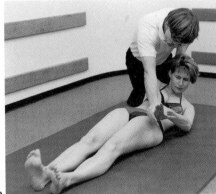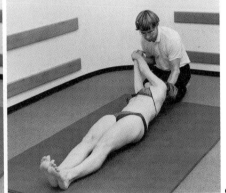

11.3.1.5 Lower Extremities

Use the patient's leg to facilitate rolling and to strengthen trunk muscles in the same way as with the arm. Flexion (anterior) patterns facilitate rolling forward, extension (posterior) patterns facilitate rolling back. The patient's head will facilitate rolling forward by going into flexion, and rolling back by going into extension.

Your distal grip is on the foot and can control the entire extremity. To make the activity effective give the principal resistance to the knee motion. Your proximal grip may be on the thigh or pelvis. When the pattern of flexion-abduction is used you may put your proximal hand on the opposite iliac crest to facilitate trunk flexion.

The commands can be specific or simple. A specific command for rolling forward using flexion-abduction is "foot up, pull your leg up, and out and roll away." The simple command is "pull your leg up." For rolling back using the pattern of extension-adduction the specific command is "push your foot down, kick your leg back, and roll back toward me." The simple command may be "kick back."

Bring the patient's leg into the elongated range of the pattern using traction to stretch the muscles of the extremity and lower trunk. Hold back on the leg motion until you see or feel the patient's trunk muscles contract then allow the leg and

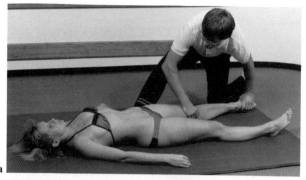

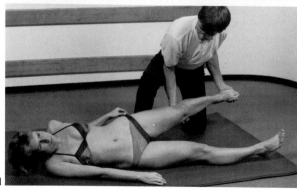

Fig. 11.7 a–f. Using one leg for rolling:
a, b forward with flexion – adduction,
c, d backward with extension – abduction,
e forward with flexion – abduction,
f backward extension – adduction

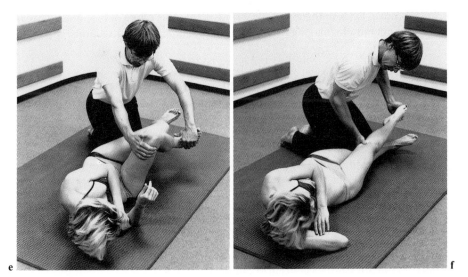

e   f

Fig. 11.7 e,f

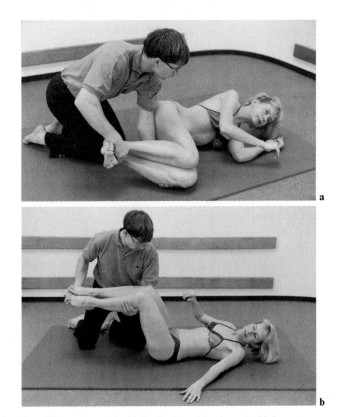

a

b

Fig. 11.8 a, b. Using both legs for rolling: *a* forward with leg flexion, *b* backward with leg extension

179

trunk to move. Lock in the leg at any strong point in its range of motion and exercise the trunk muscles and the rolling motion with repeated contractions.

The patterns are:

One leg

1 a. Flexion-adduction (Fig. 11.7 a, b): Rolling forward with trunk flexion.
1 b. Extension-abduction (Fig. 11.7 c, d): Rolling back with trunk extension and elongation.
2 a. Flexion-abduction (Fig. 11.7 e): Rolling forward with trunk lateral flexion, flexion, and rotation.
2 b. Extension-adduction (Fig. 11.7 f): Rolling back with trunk extension, elongation, and rotation.

Both legs

3 a. Lower extremity flexion (Fig. 11.8 a): Rolling forward with trunk flexion.
3 b. Lower extremity extension (Fig. 11.8 b): Rolling back with trunk extension.

11.3.2 Prone on Elbows (Forearm Support)

Lying prone on the elbows is an ideal position for working on stability of the head, neck, and shoulders. Resisted neck motions can be done effectively and without pain in this position. Resisted arm motions will strengthen not only the moving arm but also the shoulder and scapular muscles of the weight-bearing arm. The position is also a good one for exercising facial muscles and swallowing.

1. Assuming the position
   The patient can get to prone on elbows from many positions. We suggest three methods:
   – From side-sitting
   – Rolling over from a supine position
   – From a prone position (Fig. 11.9 a–d)

Resist the patient's concentric contractions if they move against gravity into the position (e. g., moving from prone to prone on elbows, Fig. 11.9 c, d). Resist eccentric control if the motion is gravity assisted (e. g., moving from side-sitting to prone on elbows).

Fig. 11.9 a–e. Prone on the elbows: *a, b* from a prone position with an arm pattern, *c, d* facilitation on the scapula, *e* stabilization

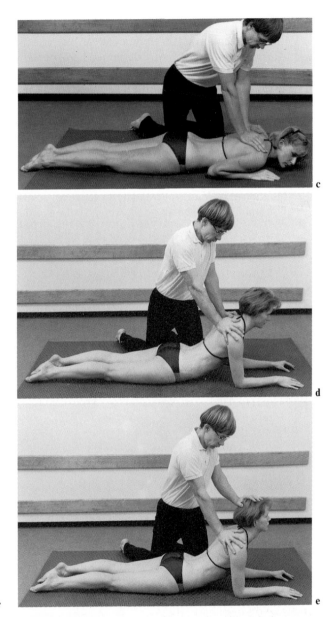

Fig. 11.9 c–e

2. Stabilizing

When the patient is secure in the position, begin stabilization with approximation through the scapula and resistance in diagonal and rotary directions. It is important that patients maintain their scapulae in a functional position and do not allow their trunk to sag. With the head and neck aligned with the trunk, give gentle resistance at the head for stabilization (Fig. 11.9 e). Rhythmic stabilization works well here. For those patients who cannot do an isometric contraction, use stabilizing reversals.

3. Motion

With the patient prone on elbows you can exercise the head, neck, upper trunk, and arms. A few exercises are described here, but let your imagination help you discover others:

- Head and neck motion: resist flexion, extension, and rotation. Try slow reversals and combination of isotonics.
- Upper trunk rotation: combine this motion with head and neck rotation. Use slow reversals and resist at the scapula or scapula and head (Fig. 11.10 a, b).
- Weight shift: shift weight completely to one arm. Combine combination of isotonics with slow reversals.
- Arm motion: after the weight shift, resist any pattern of the free arm. Use combination of isotonics followed by an active reversal to the antagonistic pattern. Use stabilizing reversals on the weight-bearing side (Fig. 11.10 c, d).

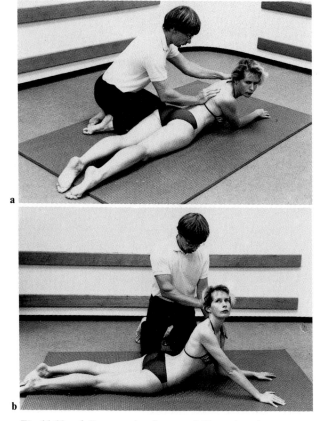

Fig. 11.10 a–d. Prone on the elbows: stability and motion

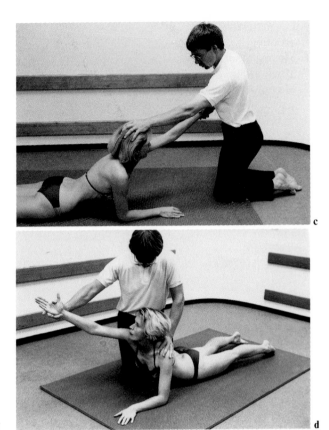

Fig. 11.10 c, d   **d**

11.3.3 Side-Sitting

This is an intermediate position between lying down and sitting. There is weight-bearing through the arm, leg, and trunk on one side. The other arm can be used for support or for functional activities. For function the patient should learn mobility in this position (scooting).

This is a good position for exercising scapular and pelvic patterns. Movement in reciprocal scapular and pelvic combinations promotes trunk mobility. Stabilizing contractions of the reciprocal patterns promote trunk stability.

We list below some of the usual activities, but do not restrict yourself only to those given: Let your imagination guide you.

1. Assuming the position
   – From side-lying
   – From prone on elbows (Fig. 11.11)
   – From sitting
   – From the crawling position
2. Balancing
3. Scapula and pelvic motions: anterior elevation and posterior depression (Fig. 11.12)

4. Leg patterns
5. Scooting
6. Moving to sitting
7. Moving to prone on elbows
8. Moving to the crawling position

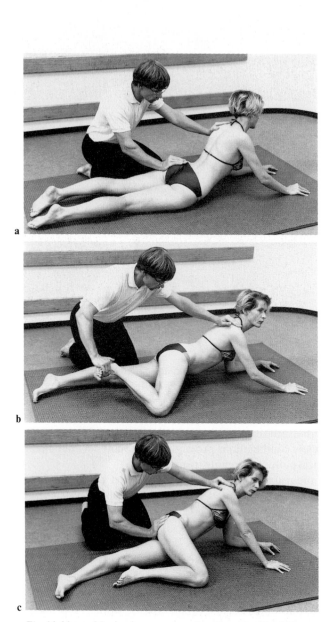

Fig. 11.11 a–e. Moving from prone on the elbows to side-sitting

184

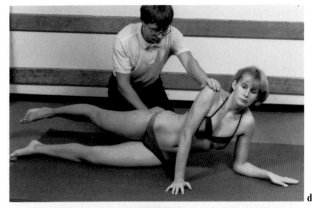

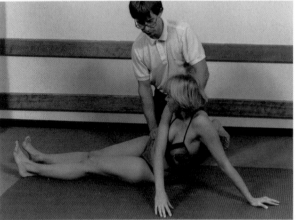

Fig. 11.11 d, e

e

d

Fig. 11.12 a

Fig. 11.12 a–c. Side-sitting: pelvis and scapula motions

11.3.4 Crawling

In a crawling position patients can exercise their trunk, hips, knees, and shoulders. Gaining the ability to move on the floor is a functional reason for activity in this position. Be sure that the scapular muscles are strong enough to support the weight of the upper trunk and that there is no knee pain.

Use the techniques stabilizing reversals and rhythmic stabilization to gain stability in the trunk and extremity joints. Resist rocking motions in all directions using combination of isotonics or dynamic reversals (slow reversals) to exercise the extremities with weight-bearing. Resisted crawling enhances the patient's ability to combine motion with stability.

1. Assuming the position
   – From prone on elbows (Fig. 11.13)
   – From side-sitting
2. Balancing (Fig. 11.14)
3. Trunk exercise (Fig. 11.15)
4. Rocking forward and back (Fig. 11.16)

5. Arm and leg exercises (Fig. 11.17)
6. Crawling
   – Resistance at scapula or pelvis
   – Resistance to leg motions (Fig. 11.18)

186

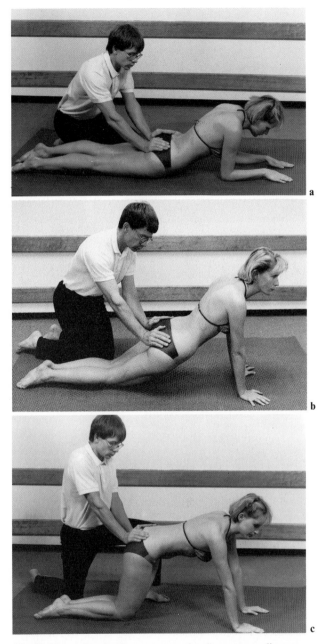

Fig. 11.13 a–c. Moving from prone on the elbows to crawling

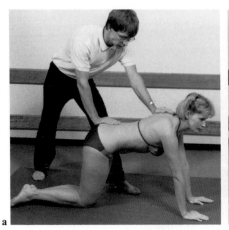

Fig. 11.14 a, b. Balancing in the crawling position

Fig. 11.15. Crawling: trunk exercise

Fig. 11.16 a

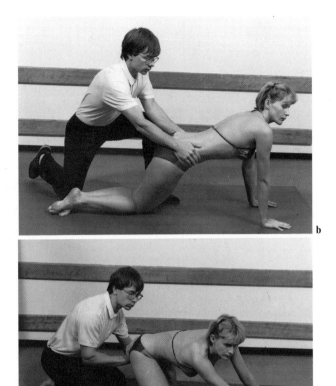

Fig. 11.16 a–c. Crawling: rocking forward and backward

Fig. 11.17 a

189

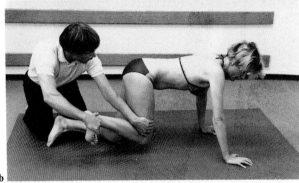

Fig. 11.17 a, b. Crawling: arm and leg exercises

Fig. 11.18 a, b. Crawling: resisting leg motions

11.3.5 Kneeling

In a kneeling position patients exercises their trunk, hips and knees, while the arms are free or used for support. For function patients go from the kneeling position to standing, or move on the floor to a piece of furniture, such as a bed or sofa.

To increase trunk strength and stability resist at the scapula and pelvis using stabilizing reversals or rhythmic stabilization. To increase the strength and range of motion of the hips and knees resist the patient's moving between kneeling and a side-sitting position. Combination of isotonics exercises the concentric and eccentric muscular functions.

1. Assuming the position
   - From a kneeling-down or side-sitting position (Fig. 11.19)
   - From a crawling position (Fig. 11.20)

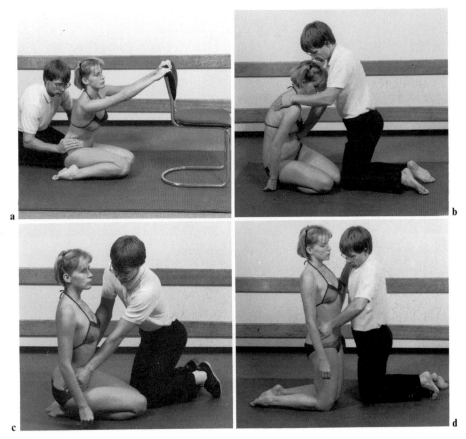

Fig. 11.19 a–d

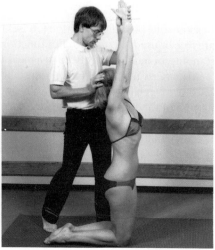

Fig. 11.19 a–f. Assuming the kneeling position

Fig. 11.20 a, b. Moving from the crawling position to kneeling

2. Balancing
   - Resistance at the scapula (Fig. 11.21 a, b)
   - Resistance at the pelvis
   - Resistance at pelvis and scapula (Fig. 11.21 c)
3. Walking on the knees
   - Forward (Fig. 11.22 a, b)
   - Backward (Fig. 11.22 c)
   - Sideways (Fig. 11.22 d, e)

192

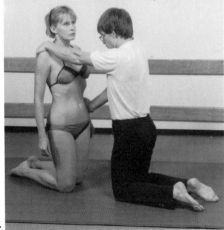

Fig. 11.21 a–c. Kneeling: stabilization with
a, b resistance at the scapula,
c resistance at the pelvis and scapula

Fig. 11.22 a, b

193

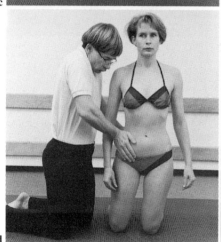

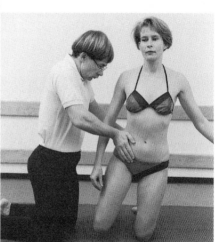

Fig. 11.22 a–e. Walking on the knees: *a, b* forward, *c* backward, *d, e* sideways

c

d

e

11.3.6 Half-Kneeling

This position is the last in the kneeling to standing sequence. For complete position the patients should move into this position with either leg forward. Use both stabilizing and moving techniques to strengthen trunk and lower extremity muscles. Shifting the weight forward over the front foot promotes an increase in ankle dorsiflexion range.

1. Assuming the position
   - From a kneeling position (Fig. 11.23)
   - From Standing
2. Balancing (Fig. 11.24)
3. Weight shift to front leg
4. Standing up (Fig. 11.25)

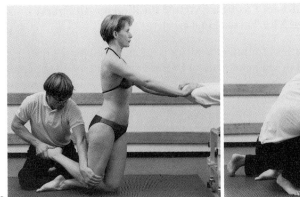

Fig. 11.23 a, b. Moving from kneeling to half-kneeling

Fig. 11.24 a, b. Balancing in half-kneeling

Fig. 11.25 a, b. Standing up from half-kneeling

11.3.7 From a Hands-and-Feet Position (Arched Position on All Fours) to a Standing Position and Vice Versa (Fig. 11.26)

The people who use this activity for function are most often those whose knees are maintained in extension. For example, patients wearing bilateral long leg braces (KAFOs) or those with bilateral above-knee prostheses can go from standing to the floor or from the floor to standing. Use of this position requires full hamstring muscle length.

a   b

Fig. 11.26 a, b. Moving to the floor and back up again

Fig. 11.27 a

196

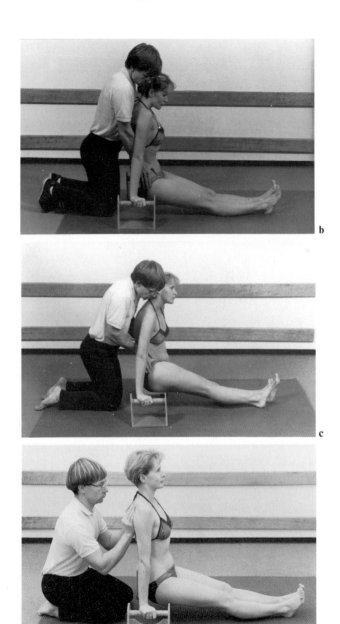

Fig. 11.27 b–d

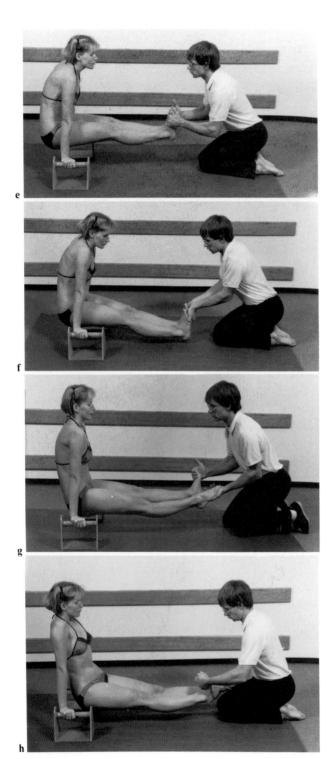

e

f

g

h

Fig. 11.27 e–h

Fig. 11.27 a–i. Exercises in long-sitting: *a* stabilization, *b–d* pushing up, *e–h* providing resistance to the legs, *i* scooting forward

11.3.8 Exercise in a Sitting Position

11.3.8.1 Lon g-Sitting

This position is functional for bed activities such as eating and dressing. Exercising in long-sitting is used to increase the patient's balance. Because the patient can sit on the floor mats, this is a safe position for independent balance work. Long-sitting is a good position for exercises to increase arm and trunk strength. Patients can practice the lifts used for transfers.

1. Assuming the position
   - From side-sitting
   - From supine
2. Balancing with and without upper extremity support (Fig. 11.27 a)
3. Pushing up exercises
   - Resistance at the pelvis and shoulders (Fig. 11.27 b–d)
   - Resistance at the legs (Fig. 11.27 e–h)
4. Scooting forward (Fig. 11.27 i) and backward

11.3.8.2 Short-Sitting

To use their arms for other activities, patients need as much trunk control as possible. To reach for distant objects they need to combine trunk stability with trunk, hip, and arm motion. Patients need to exercise while sitting on the side of the bed and in a chair as well as on the mats. Static exercises in short-sitting will increase the patient's trunk and hip stability. Dynamic exercises will increase trunk and hip motion. Resisting the patient's strong arms will provide irradiation to facilitate weaker trunk and hip muscles.

1. Assuming the position from side-lying (Fig. 11.28)
   Resist the patient's concentric contractions while they move into sitting. Resist the eccentric control as they lie down.
2. Balancing
   Use stabilizing reversals or rhythmic stabilization to increase trunk stability. Resist at the shoulders, pelvis, and head (Fig. 11.29).

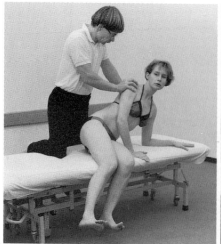
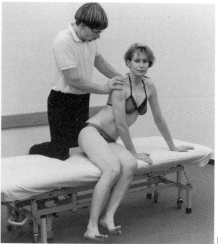

Fig. 11.28 a, b. Moving from sidelying to short-sitting

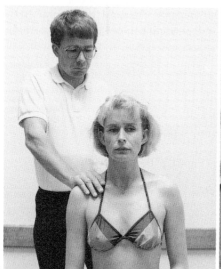
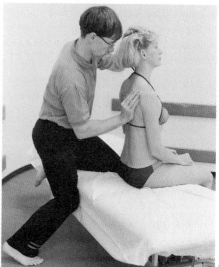

Fig. 11.29 a, b. Stabilization in short-sitting

- With and without upper extremity support
- With and without lower extremity support

3. Trunk exercises

Use dynamic reversals (slow reversals) and combination of isotonics to increase the patient's trunk strength and coordination. Resist at the shoulders (Fig. 11.30a) or use chopping (Fig. 11.30b) and lifting combinations to get added irradiation.

- Trunk flexion (Fig. 11.30c) and extension (Fig. 11.30d)

200

– Reaching forward and to the side with return: this requires hip flexion, extension, lateral motion, and rotation with the trunk remaining stable

4. Moving

These activities teach mobility in sitting and exercise pelvic and hip muscles.

– Forward and back (Fig. 11.30e)
– From side to side

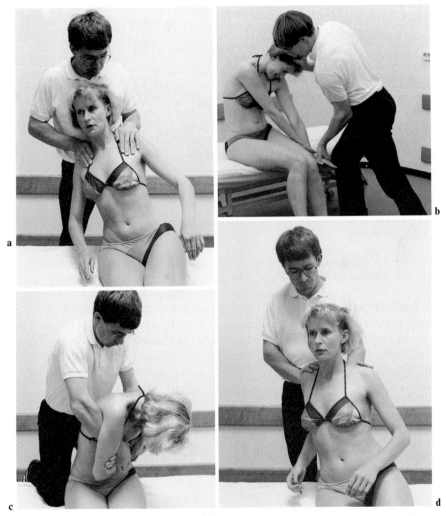

Fig. 11.30 a–e. Short-sitting: *a* dynamic trunk exercise, *b* chopping, *c* trunk flexion, *d* trunk extension, *e* moving forward

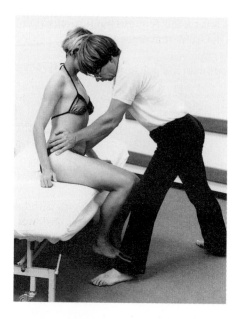

Fig. 11.30 e

11.3.9 Bridging

In the hook-lying position the patient exercises with weight-bearing through the feet but without danger of falling. Lifting the pelvis from the supporting surface makes it easier for a person to move and dress in bed.

Working in the hook-lying position requires some selective control of the leg muscles. Patients must keep their knees flexed while extending their hips and pushing with their feet. When patients push against the mats with their arms, their upper trunk, neck, and upper extremity muscles are exercised. Resist concentric, eccentric, and stabilizing contractions to increase strength and stability in the trunk and lower extremity.

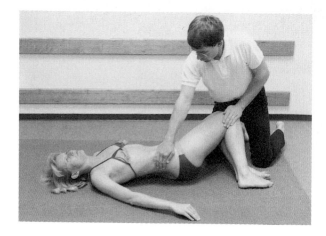

Fig. 11.31. Lower trunk rotation in the hook-lying position

1. Assuming the position
2. Stabilizing
   – With approximation into the pelvis
   – With approximation into the feet
   – Without approximation

   Use resistance with approximation at the legs to facilitate lower extremity and trunk stability. Give the resistance in all directions. Resistance in diagonals will recruit more trunk muscle activity. As the patient gains strength, decrease the amount of approximation. Resist the legs together and separately. Resist both legs in the same direction and in opposite directions when working them separately.

3. Lower trunk rotation in the hook-lying position

   The motion begins with the legs moving down diagonally (distally) toward the floor. When the hips have completed their rotation the pelvis rotates, followed by the spine. The return to upright requires a reverse timing of the motion. The lumbar spine must derotate first, then the pelvis, and then the legs. Correct timing of this activity is important. Limit the distance that the legs descend to the ability of the patient to control the motion. Figure 11.31 shows resistance to returning to the upright leg position with lower trunk rotation to the right. You can use combination of isotonics and slow reversals to teach and strengthen this activity.

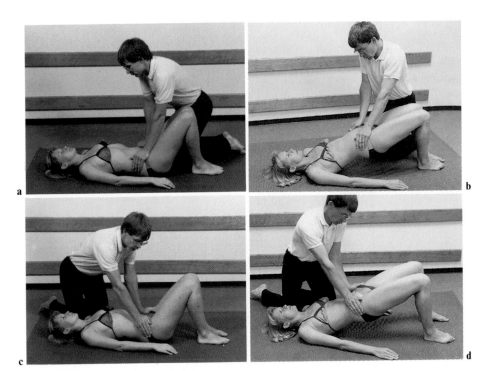

Fig. 11.32 a–d. Bridging on two legs in supine position

4. Bridging
   – Stabilize the pelvis with resistance in all directions (Fig. 11.32 a, b, resisting from below; Fig. 11.32 c, d, resisting from above.)
   – Lead with one side of the pelvis
   – Resist static and dynamic rotation of the pelvis
   – Scoot the pelvis from side to side

   Use combination of isotonics to strengthen the patient's antigravity control. *Caution:* Monitor and control the position of the patient's lumbar spine while the pelvis is elevated.

5. Other bridging activities
   – Stepping in place
   – Walking the feet: apart, together, to the side, away from the body (into extension) and back
   – Bridge on one leg (Fig. 11.33)
   – Bridge while bearing weight on the arms (Figs. 11.34–11.36).

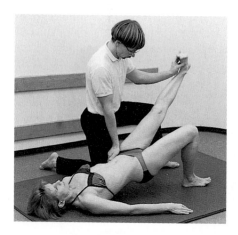

Fig. 11.33. Bridging on one leg

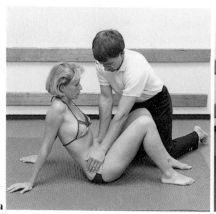

Fig. 11.34 a, b. Bridging on the hands

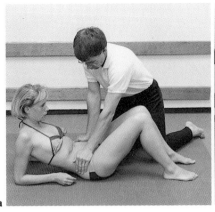

Fig. 11.35 a, b. Bridging on the elbows

Fig. 11.36 a, b. Bridging on the arms and one leg

Reference

VanSant AF (1991) Life-span motor development. In: Contemporary management of motor control problems. Proceedings of the II Step conference. Foundation for Physical Therapy, Alexandria, VA

Further Reading

Portney LG, Sullivan PE, Schunk MC (1982) The EMG activity of trunk-lower extremity muscles in bilateral-unilateral bridging. Phys Ther 62: 664

Schunk MC (1982) Electromyographic study of the peroneus longus muscle during bridging activities. Phys Ther 62: 970–975

Sullivan PE, Portney LG, Rich CH, Langham TA (1982) The EMG activity of trunk and hip musculature during unresisted and resisted bridging. Phys Ther 62: 662

Sullivan PE, Portney LG, Troy L, Markos PD (1982) The EMG activity of knee muscles during bridging with resistance applied at three joints. Phys Ther 62: 648

Troy L, Markos PD, Sullivan PE, Portney, LG (1982) The EMG activity of knee muscles during bilateral-unilateral bridging at three knee angles. Phys Ther 62: 662

12 Gait Training

12.1 Introduction

Walking is a major goal for most patients. Effective walking requires the ability to change direction and to walk backward and sideways as well as forward. Being able to go up and down curbs, climb stairs and hills, and open and close doors increases the utility of the activity. To be totally functional the individual should be able to get down onto the ground and back up to standing again.

Walking must be so automatic that the person can turn his or her attention to the requirements of the environment, such as traffic, while continuing to walk. For walking to be safe the individual must be able recover balance if it is disturbed, either by the act of walking or by outside forces. To walk more than a few steps requires gait that is as energy efficient as possible. To get around the house in a reasonable time requires less energy and speed than to walk around a supermarket or cross a street. The individual needs enough endurance and skill to walk the necessary distances at a practical speed (Lerner-Frankiel et al. 1986).

For an easy way to evaluate a person's gait, place your hands on his or her pelvis and feel what is happening as the person walks normally. Your hands go on the iliac crest as though you were resisting pelvic elevation. When evaluating, do not resist or approximate, just feel.

Standing up is both a functional activity and a first stage in walking. The person should be able to stand up and sit down on surfaces of different heights. Although everyone varies in the way he or she gets from sitting to standing, the general motions can be summarized as follows (Nuzik et al. 1986):

1. The first part of the activity:
   a) The head, neck, and trunk move into flexion.
   b) The pelvis moves into a relative anterior tilt.
   c) The knees begin to extend and move forward over the base of support.
2. The last part:
   a) The head, neck, and trunk extend back toward a vertical position.
   b) The pelvis goes from an anterior to a posterior tilt.
   c) The knees continue extending and move backward as the trunk comes over the base of support.

Until studies bring other information, we assume that sitting down involves the reverse of these motions. Control comes from eccentric contraction of the muscles used for standing up.

Adequate joint range of motion in the hip, knee, and ankle is needed for standing and walking to be practical. Limitation of motion at these joints, imposed by joint restrictions or by orthoses, will interfere with the normal swing and stance and decrease walking efficiency (Murray et al. 1964).

The individual needs strength in the muscles of the ankle, knee, hip, and trunk to stand up and walk without external support. Correct timing of the contraction and relaxation in these muscle groups is required for practical balance and gait (Horak and Nashner 1986; Eberhart et al. 1954). Exercises on the mats and treatment table are used to help bring the muscles to the level of strength needed for function.

When describing the act of walking it is usual to speak of a complete gait cycle. One cycle lasts from the heel strike of one leg to the next heel of strike of that same leg. Regular walking has one period when both feet are on the ground (double limb support) and one period when only one foot is on the ground (single limb support; Inman et al. 1981). Each gait cycle is divided into two phases (Perry 1967):

1. *Stance phase:* The foot is on the ground supporting the body.
   a) Heel strike to foot flat: Weight is being accepted on the leg.
   b) *Mid-stance:* The trunk moves forward over the flat foot.
   c) *Push-off:* The body is ahead of foot. The heel is just beginning to rise.
   d) *Balance assistance:* The toe is still touching the floor with the knee passively flexed. The leg still assists balance.

2. *Swing phase:* The foot is off the ground, the leg is advancing and preparing for the next support phase.
   a) *Toe-off:* The foot lifts from ground with active hip and knee flexion. The leg begins to swing forward.
   b) *Reach:* The leg swings forward. The knee extends by pendulum effect while the ankle dorsiflexor muscles are active to keep the toe from dragging.

12.2 The Theory of Gait Training

We use all the basic procedures and many of the techniques when working with our patients in standing and walking. Resistance, appropriately used, increases the patient's ability to balance and move. When the strong motions are resisted in standing and walking, irradiation will cause contraction of weaker trunk and lower extremity muscles. These weaker muscles will contract whether or not braces or other supports are used.

As the patient's ability increases he or she must be allowed and encouraged to stand and walk as independently as possible. During these practice sessions, no verbal or physical cuing should be given, and only the assistance necessary for safety. Allow patients to solve problems and correct mistakes on their own. Alternate resisted gait training with independent walking during a treatment session. After an activity is mastered, resisted work is used for strengthening.

Resisted gait activities can be used to treat specific joint and muscle dysfunctions in the upper and lower extremities. For example, exercise the lateral and medial ankle muscles by resisting sideways stepping. The shoulder, elbow, wrist, and hand are exercised when the patient holds the parallel bar while balancing or moving against resistance.

12.3 The Procedures of Gait Training

The primary emphasis in gait training is on the patient's trunk. Approximation through the pelvis during stance and stretch to the pelvis during swing facilitate the muscles of the lower extremities and the trunk (S. S. Adler, unpublished, 1976). Proper placement of the hands allows the therapist to control the position of the patient's pelvis, moving it toward an anterior or posterior tilt as needed.

Resistance to balance and motion is most effective when given in a diagonal direction. The therapist controls the direction of resistance by standing in the chosen diagonal. The therapist's *body position* also allows the use of body weight for approximation and resistance.

Resisted gait activities are exaggerations of normal motions. Large-amplitude body motions are resisted during weight shifting. During walking the pelvic motions are larger and the steps are higher. Resistance to the large motions helps the patient gain the strength and skill needed to stand and walk functionally.

12.3.1 Approximation and Stretch

12.3.1.1 Approximation
To approximate, place the heel (carpal ridge) of each hand on the anterior crest of the ilium, above the anterior superior iliac spine (ASIS). Your fingers point down and back in the direction of the force. Keep the patient's pelvis in a slight posterior

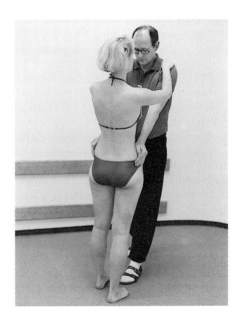

Fig. 12.1. Correct way of giving facilitation on the pelvis for approximation and stretch

tilt. The direction of the approximation force should go through the ischial tuberosities towards the patient's heels. Apply the approximation sharply and maintain it while adding resistance.

The following precautions should be taken:

1. Your wrists should be extended only a few degrees past neutral to avoid wrist injury.
2. To avoid fatigue and shoulder pain, use your body weight to give the approximation force. Keep your elbows in a near-extended position so the body weight can come down through your arms (Fig. 12.1).
3. Your hands should stay on the crest of the ilium to prevent pain and bruising of the tissues of the abdomen and ASIS.

12.3.1.2 Stretch

To give stretch at the pelvis, use the same grip as used for approximation. When the patient's foot is unweighted, stretch the pelvis down and back. The stretch is the same as the one used for the pattern of anterior elevation of the pelvis.

Precautions to be taken are:

1. Your hands should stay on the crest of the ilium and not slip down to the ASIS.
2. The stretch should move the pelvis down and back. Do not rotate the patient's body around the stance foot.

12.3.2 Using Approximation and Stretch

Standing

Use approximation to facilitate balance and weight-bearing. Give resistance immediately to the resulting muscle contractions. The direction of the resistance determines which muscles are emphasized:

1. Resistance directed diagonally backward facilitates and strengthens the anterior trunk and limb muscles.
2. Resistance directed diagonally forward facilitates and strengthens the posterior trunk and limb muscles.
3. Rotational resistance facilitates and strengthens all the trunk and limb muscles with an emphasis on their rotational component.

Approximation with resistance through the shoulder girdle places more demand on the upper trunk muscles. Put your hands on the top of the shoulder girdle to give the approximation. Be sure that the patient's spine is properly aligned before giving any downward pressure.

The walking descriptions below apply to walking forward with the therapist in front of the patient. The same principles hold true when the patient walks backward or sideways. When the patient walks backward, stand behind the patient and direct your pressure down and forward. For sideways balancing and walking stand to the side of the patient and direct your pressure down and laterally.

Swing

Stretch and resistance to the upward and forward motion of the pelvis promotes both the pelvic motion and the hip flexion needed for swing. You can further facilitate hip flexion with *timing for emphasis*. Do this by blocking the pelvic motion until the hip begins to flex and the leg to swing forward.

Stance

Approximation combined with resistance to the forward motion of the pelvis facilitates and strengthens the extensor musculature. Give approximation in a downward and backward direction to the stance leg at or just after heel strike to promote weight acceptance. Approximate again at any time during the stance phase to maintain proper weight-bearing.

12.4 Practical Gait Training

12.4.1 Preparatory Phase

A necessary part of a patient's gait training is learning to manage a wheelchair. These activities are a part of both gait and daily living training. Use all the basic procedures to help the patient gain skill in these activities. Repetition combined with resistance enables the patient to master the activities in the shortest possible time.

12.4.1.1 Managing the Wheelchair

Guiding and resisting the motions used to manage the wheelchair will help the patient master these activities. The general activities are:

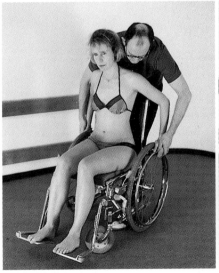

a   b

Fig. 12.2 a, b

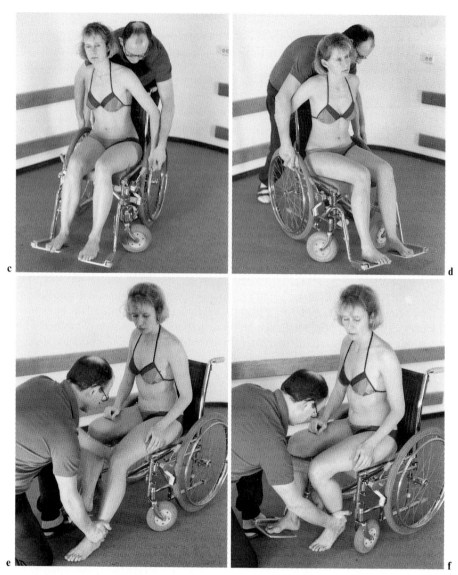

Fig. 12.2 a–f. Managing the wheelchair: ***a, b*** wheeling forward, ***c, d*** wheeling backward, ***e, f*** wheeling forward with resistance on the leg

1. Wheeling the chair
   - Forward (Fig. 12.2 a, b) and backward (Fig. 12.2 c, d) with resistance to the arms
   - Forward with resistance to the leg (Fig. 12.2 e, f)
2. Locking and unlocking the brakes (Fig. 12.3)
3. Removing and replacing the arm rests (Fig. 12.4)
4. Managing the foot pedals (Fig. 12.5)

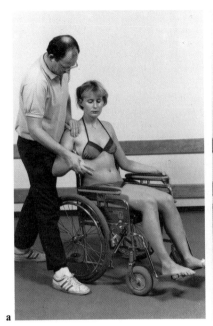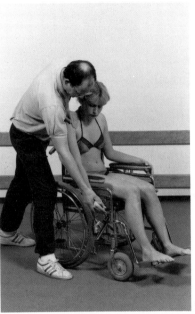

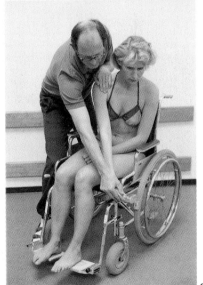

Fig. 12.3 a–c. Managing the brakes

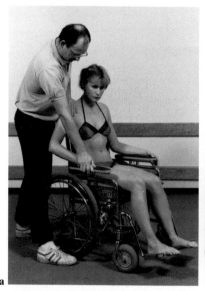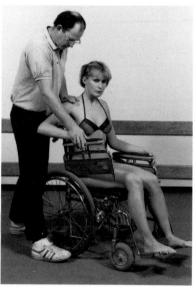

Fig. 12.4 a, b. Managing the armrests

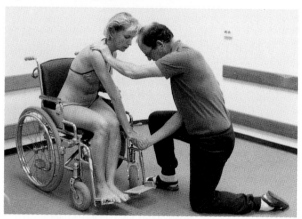

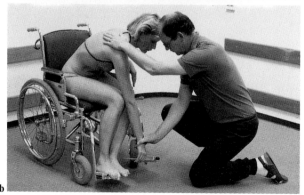

Fig. 12.5 a, b. Managing the footrests

12.4.1.2 Sitting

It is necessary for the patient to be able to sit upright and move in a chair. Stretch and resistance at the pelvis can guide the patient into the proper erect posture with ischial weight-bearing. Resistance at the scapula and head teaches and strengthens trunk stability. Use stretch and resistance to the appropriate pelvic motions to teach the patient to move forward and backward in the chair. While working on these activities, evaluate the patient's strength and mobility. Treat any problems that limit function and reevaluate in a sitting position after treatment.

Example: You cannot get the patient's pelvis positioned for proper ischial weight-bearing. You think this may be due to limitation in range of pelvic motion.

Put the patient on mats and assess pelvic motion using pelvic patterns. Treat any limitations in range and strength with exercises for the pelvic and scapular patterns. After treatment, take the patient back into the wheelchair and reevaluate the pelvic position in a sitting position.

12.4.1.3 Sitting Activities

Getting into the Upright Sitting Position

- Use combination of isotonics with resistance at the head and shoulders to get the upper trunk into an erect position (Fig. 12.6).
- Use rhythmic initiation and stretch at the pelvis to achieve an anterior tilt.

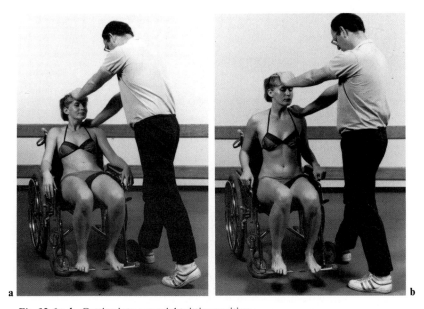

a   b

Fig. 12.6 a, b. Getting into an upright sitting position

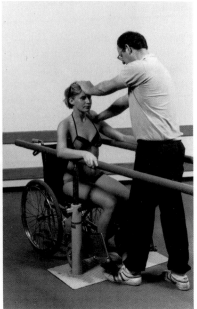

Fig. 12.7 a, b. Stabilizing the sitting position

Stabilizing in the Upright Position

Use stabilizing reversals (Fig. 12.7 a, resistance at pelvis and scapula; Fig. 12.7 b resistance at head and scapula)

- At the head
- At the shoulders
- At the pelvis
- A combination of all of them

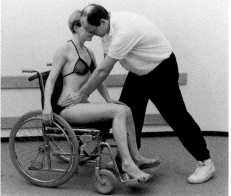

Fig. 12.8 a, b. Moving forward in the chair

216

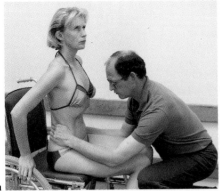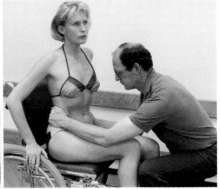

Fig. 12.9 a, b. Moving backward in the chair

Moving in the Chair
Use repeated stretch, rhythmic initiation, and isotonic reversals (slow reversals)

- Pelvic anterior elevation for moving forward (Fig. 12.8)
- Pelvic posterior elevation for moving backward (Fig. 12.9, chair arm removed to show pelvic movement).

The following sections are an artificial grouping of activities. A treatment usually proceeds smoothly through all activities in a functional progression. The patient moves forward in the chair, stands up, gets his or her balance, and walks. You break down the activities as needed and work on those which are not yet functional or smooth.

12.4.2 Standing Up and Sitting Down

To increase the patient's ability to stand up, place your hands on the patient's iliac crests (Fig. 12.10 a), rock or stretch the pelvis into a posterior tilt, and resist or assist as it moves into an anterior tilt. Rhythmic initiation works well with this activity. Three repetitions of the motion are usually enough. On the third repetition give the command to stand up. Guide the pelvis up and into an anterior tilt as the patient moves toward standing. Assist the motion if that is needed, but resist when the patient can accomplish the act without help. As soon as the patient is upright guide the pelvis into the proper amount of posterior tilt. Approximate through the pelvis to promote weight bearing.

Getting to Standing
- Moving forward in chair: The same as practice in sitting.
- Placing hands: Use rhythmic initiation to teach patients where to put their hands. Use stabilizing contractions and combination of isotonics to teach them how to assist with their arms.
- Using the parallel bars
- Using the chair arms

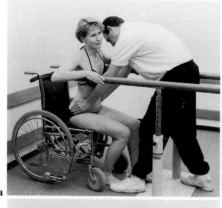
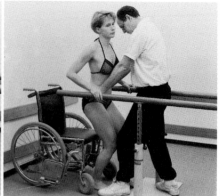

Fig. 12.10 a–c. Getting to a standing position

- Rocking the pelvis: Use rhythmic initiation and stretch to get the pelvis tilted forward (Fig. 12.10 a).
- Coming to standing: Guide and resist at the pelvis (Fig. 12.10).
  Guide and resist at the shoulders if the patient cannot keep the upper trunk in proper alignment.

Sitting Down
- Placing hands to assist: Use the same techniques as in standing up.
- Sitting down: Use resistance at the pelvis or pelvis and shoulders for eccentric control. When the patient is able, use combination of isotonics by having the patient stop part way down and then stand again.

12.4.3 Standing

Stand in a diagonal in front of the leg that is to take the patient's weight initially. Guide the patient to that side and use approximation and stabilizing resistance at the pelvis to promote weight-bearing on that leg. If weight is to be born equally on both legs, stand directly in front of the patient.

218

Weight Acceptance

– Combine approximation through the pelvis on the strong side with stabilizing resistance at the pelvis.
– Combine approximation through the pelvis on the weaker side (knee blocked if
  necessary) with stabilizing resistance at the pelvis.

Stabilization

– Combine approximation and stabilizing reversals at the pelvis for the lower
  trunk and legs (Fig. 12.11 a).
– Combine approximation and stabilizing reversals at the shoulders for the upper
  and lower trunk (Fig. 12.11 b).
– Using combination of isotonics with small motions or stabilizing reversals, resist
  balance in all directions. Work at the head, the shoulders, the pelvis, and combination of these.

One-Leg Standing

Use this activity to promote weight-bearing in stance and to facilitate pelvic and
hip motion in swing. The patient stands on one leg with the other hip flexed.
The flexed hip should be above 90°, if possible, to facilitate hip extension on the
other leg. If the patient is not able to hold up the flexed leg, assist by placing the
patient's knee above your pelvis and giving a compressive force to hold the leg
in place (Fig. 12.12 a). Alternate the weight-bearing leg frequently to avoid fatigue.

Fig. 12.11 a, b. Stabilization at the pelvis and on the shoulders

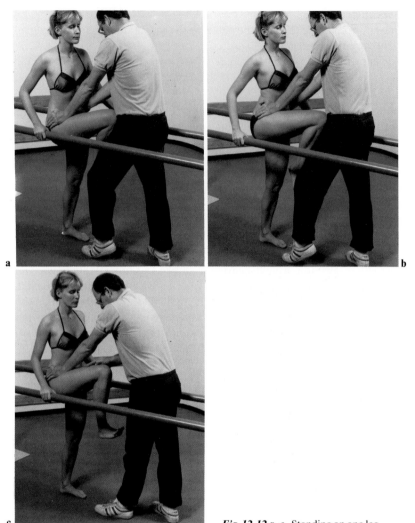

a

b

c   *Fig. 12.12 a–c.* Standing on one leg

Emphasis on stance leg:
- Approximate through the pelvis to encourage weight-bearing (Fig. 12.12 c).
- Use combination of isotonics with small motions of stabilizing reversals at the pelvis to resist balance in all directions.

Emphasis on swing leg:
- Use repeated stretch with resistance to facilitate anterior elevation of the pelvis on that side (Fig. 12.12 a, b).
- Use combination of isotonics to facilitate hip flexion.

Weight Shifting

Use this activity both as a preliminary to stepping and to exercise specific motions in the lower extremity. Exaggerated weight shift forward or laterally exercises the hip hyperextension and lateral motions, knee stability, and ankle motion.

Start the weight shift activity by stabilizing the patient on one leg. Then resist as he or she shifts weight to the other leg. Using approximation and resistance, stabilize the patient in the new position. You can complete this exercise in one of two ways:

a   b   c   d

Fig. 12.13 a–d. Shifting the weight forward, stepping forward

221

1. By resisting eccentric contraction as the patient allows you to push him or her slowly back over the other leg.
2. By resisting concentric contractions while the patient actively shifts weight to the other leg. In this case, you must move your hands to give resistance to the motion.

Weight Shift from Side to Side
- Stabilizing resistance
- Resistance to sideways weight shift
- Approximation and resistance on weight-bearing side
- Resisted eccentric or concentric return:
  Eccentric: keep your hands positioned to resist the original weight shift
  Concentric: move your hands to the opposite side of the pelvis; resist an antagonistic weight shift

Weight Shift Forward and Backward (Stride Position) (Fig. 12.13)

When working on this activity it is important for the patient to shift the whole pelvis and trunk forward and backward. Do not allow the patient to come forward in a sideways position. Stand in front of the patient to emphasize forward weight shift, and behind to emphasize backward weight shift. As always, stand in the line of the patient's motion. The example below is for shifting forward; reverse the directions for shifting backward.

Example: The patient is standing on his right leg with the left leg in front. You stand in a stride diagonally in front of the left leg with your left foot forward in front of his right foot. Your weight is on your forward foot.
- *Stabilize:* Use approximation and resistance to stabilize the patient on the back leg.
- *Resist:* Give diagonal resistance as the patient's weight shifts from the back to the front leg. Let the patient's movement push you back over your right leg.
- *Stabilize:* Give approximation through the left (front) leg combined with bilateral resistance to stabilize the patient on the front leg. Use your body weight to give the resistance.
- *Resist:* Give diagonal resistance to eccentric or concentric work to return the patient's weight to his back leg:
  Eccentric: keep your hands positioned on the anterior superior iliac crests.
  Concetric: move your hands to the posterior superior iliac crests.

Repeated Stepping (Forward and Backward)
This activity goes with weight shifting. You may have the patient shift weight three or four times before stepping or ask for a step following each weight shift. As the patient steps, you shift your body to place it in the line of the new stance leg. Use this activity to exercise any part of swing or stance that needs work. You may modify the activity to do repeated stepping sideways.

Example: Repeated stepping forward and back with the right leg.
- *Stabilize* on the back leg.
- *Resist* the weight shift to the forward (left) leg.
- *Stabilize* on the forward leg.

- *Stretch and resist:* When the patient's weight is on the left leg, stretch the right side of the pelvis down and back. Resist the upward and forward motion of the pelvis to facilitate the forward step of the right leg. As the patient steps with the right leg, you step back with your left leg and put your weight on it.
- *Stabilize* on the forward leg.
- *Resist* the weight shift back to the left leg:
  Eccentric: maintain the same grip as you push the patient slowly back over the left leg.
  Concentric: shift your grip to the posterior pelvic crest and resist the patient shifting his or her weight back over the left leg.
- *Resist* a backward step with the right leg:
  Eccentric: tell the patient to step back slowly while you maintain the same grip and try to push the pelvis and leg back rapidly.
  Concentric: shift your grip to the posterior pelvic crest, then stretch and resist an upward and backward pelvic motion to facilitate a backward step with the right leg.

12.4.4 Walking

After weight shifting and repeated stepping, it is time to put all the parts together and let the patient walk. When the objective of the walk is evaluation or function, give the patient just enough support to maintain safety. When the objective is to strengthen and reeducate, use approximation, stretch, and resistance as you did with weight shift and repeated stepping. *Caution:* Resisted walking interrupts the patient's momentum and decreases velocity.

12.4.4.1 Forward

Standing in Front of the Patient
Mirror the patient's steps. As the patient steps forward with the right leg you step back with your left. Use the same procedures and techniques as you used for repeated stepping (Fig. 12.14 a, b).

Standing Behind the Patient
Both you and the patient step with the same leg. When standing behind, your fingers are on the iliac crest. Your hands and forearms form a line that points down through the ischial tuberosities towards the patient's heels. Your forearms press against the patients gluteal muscles (Fig. 12.14 c).
   Standing behind is advantageous when:

1. The patient is much taller than you are: you can use your body weight to pull down and back on the pelvis for approximation, stretch, and resistance.
2. You want to give the patient an unobstructed view forward.
3. The patient is using a walker or other walking aid.

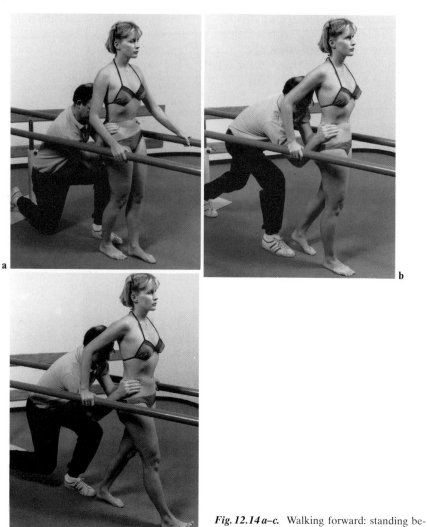

Fig. 12.14 a–c. Walking forward: standing behind the patient

12.4.4.2 Backward (Fig. 12.15)

Walking backward requires trunk control and exercises hip hyperextension in
swing. It is a necessary part of functional walking:

– Stand behind the patient. Place the heel of your hand on the posterior superior iliac crest and give pressure down and forward.
– The patient must maintain an upright trunk while walking backward.

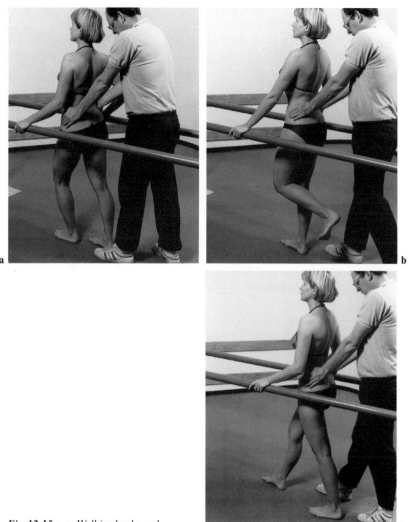

Fig. 12.15 a–c. Walking backward

12.4.4.3 **Sideways** (Fig. 12.16, walking sideways; Fig. 12.17, braiding)

Walking sideways exercises the lateral muscles of the trunk and legs: Stand so the patient walks towards you. Give approximation, stretch, and resistance through the pelvis. If the upper trunk needs stabilizing, place one hand on the lateral aspect of the shoulder.

Fig. 12.16 a, b. Walking sideways

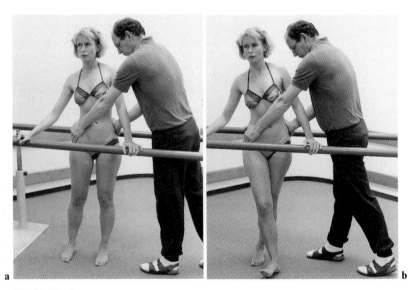

Fig. 12.17 a, b

226

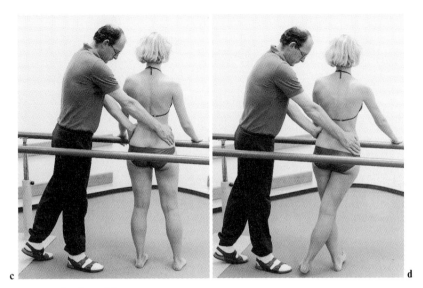

Fig. 12.17 a–d. Braiding

Fig. 12.18 a, b

227

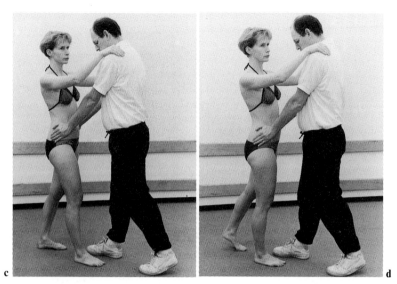

Fig. 12.18 a–d. Walking outside the bars

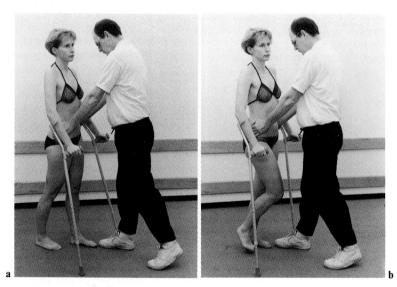

Fig. 12.19 a, b. Walking with crutches

228

12.4.5 Other Activities

Here we illustrate some other activities we consider important for the patient to master. Use the procedures and techniques that are appropriate for each situation.

- Walking outside the bars (Fig. 12.18)
- Walking with crutches (Fig. 12.19)
- Going up and down stairs (Fig. 12.20)
- Going up and down curbs (Fig. 12.21)
  Curbs are one step without a railing.
- Going down and getting up from the floor. (We cover this activity in Chap. 11, but also consider it an important part of walking.)

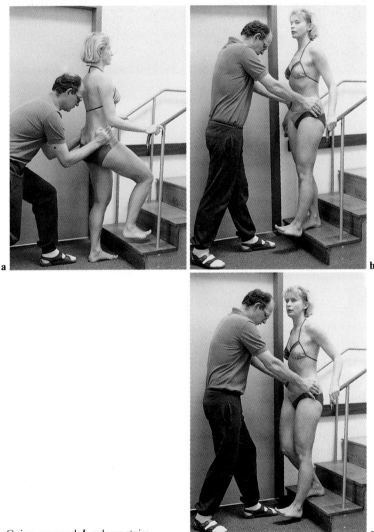

Fig. 12.20 a–c. Going *a* up and *b, c* down stairs

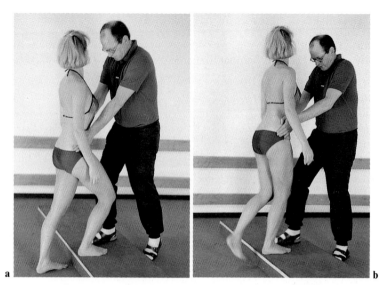

a   b

Fig. 12.21 a, b. Going up a curb

References

Adler SS (1976) Influence of "Joint Approximation" on lower extremity extensor muscles: an EMG study. unpublished thesis presented at APTA annual conference, New Orleans

Eberhart HD, Inman VT, Bresler B (1954) The principal elements in human locomotion, in Human Limbs & Their Substitutes Klopteg PE, Wilson PD (ed.). McGraw-Hill Book Co., Inc

Horak FB, Nashner LM (1986) Central programming of postural movements: adaptation to altered support-surface configurations. J Neurophysiol 55 (6): 1369–1381

Inman VT, Ralston HJ, Todd F (1981) Human Walking. Baltimore, Williams & Wilkins

Lerner-Frankiel MB, Vargas S, Brown M, Krusell, L (1986) Functional community ambulation: what are your criteria? Clinical Management 6 (2): 12–15

Murray MP, Drought AB, Kory RC (1964) Walking Patterns of Normal Men, JBJS 46–A (2): 335–360

Nuzik S, Lamb R, VanSant A, Hirt S (1986) Sit-to-stand movement pattern, a kinematic study. Phys Ther 66 (11): 1708–1713

Perry J (1967) The mechanics of walking, a clinical interpretation. in Principles of Lower-Extremity Bracing. ed: Perry J, Hislop HJ. American Physical Therapy Association, Washington, D. C.

Further Reading

Posture Control and Movement

Finley FR, Cody KA (1969) Locomotive characteristics of urban pedestrians. Arch Phys Med Rehabil 51: 423–426

Gahery Y, Massion J (1981) Co-ordination between posture and movement. Trends Neuro Sci 4: 199–202

Nashner LM (1980) Balance adjustments of humans perturbed while walking. J Neurophysiol 44: 650–664

Nashner LM (1982) Adaptation of human movement to altered environments. Trends Neuro Sci 5: 358–361

Nashner LM, Woollacott M (1979) The organization of rapid postural adjustments of standing humans: an experimental-conceptual model. In: Talbott RE, Humphrey DR (eds) Posture and movement. Raven, New York 1979

Woollacott MH, Shumway-Cook A (1990) Changes in posture control across the life span – a systems approach. Phys Ther 70: 799–807

Gait

Inman VT, Ralston HJ, Todd F (1981) Human walking. Williams and Wilkins, Baltimore

Kettelkamp DB, Johnson RJ, Schmidt GL, et al. (1970) An electrogoniometric study of knee motion in normal gait. J Bone Joint Surg [A] 52: 775–790

Mann RA, Hagy JL, White V, Liddell D (1979) The initiation of gait. J Bone Joint Surg [A] 61: 232–239

McFadyen BJ, Winter DA (1988) An integrated biomechanical analysis of normal stair ascent and descent. J Biomechan 21: 733–744

Murray MP, Kory RC, Sepic SB (1970) Walking patterns of normal women. Arch Phys Med Rehabil 51: 637–650

Murray MP, Drought AB, Kory RC (1964) Walking patterns of normal men. J Bone Joint Surg [A] 46: 335–360

Nashner LM (1976) Adapting reflexes controlling the human posture. Exp Brain Res 26: 59–72

Perry J (1992) Gait analysis, normal and pathological function. Slack, Thorofare NJ

Sutherland DH (1966) An electromyographic study of the plantar flexors of the ankle in normal walking on the level. J Bone Joint Surg [A] 48: 66–71

Sutherland DH, Cooper L, Daniel D (1980) The role of the ankle plantar flexors in normal walking. J Bone Joint Surg [A] 62: 354–363

Sutherland DH, Olshen R, Cooper L, Woo SLY (1980) The development of mature gait. J Bone Joint Surg 62: 336–353

13 Vital Functions

13.1 Introduction

Therapy for the vital functions includes exercises for the face, tongue, breathing, and swallowing. Treatment of these areas is of particular importance when facial weakness, swallowing, and respiratory difficulties are involved. You can do breathing and facial exercises at any time. They are particularly useful when a patient becomes fatigued from other activities. Use breathing exercises for relaxation if the patient is tense or in pain.

13.1.1 Stimulation and Facilitation

Use the same procedures and techniques when treating problems in breathing, swallowing, and facial motion as you do when treating other parts of the body. Quick stretch and resistance promote muscle activity and increase strength. Proper grip and pressure guide and facilitate the movements. Additional facilitation can be achieved by using ice for facilitation (quick ice): use two or three quick, short strokes with the ice on the skin overlying the muscles, on the tongue, or inside the mouth.

Use bilateral movements (both sides together) when exercising the face or chest. Contraction of the muscles on the stronger or more mobile side will facilitate and reinforce the action of the more involved muscles. Timing for emphasis, by preventing full motion on the stronger side, will further promote activity in the weaker muscles.

13.2 Facial Muscles

The muscles of the face have many functions, including facial expression, jaw motion, protecting the eyes, aiding in speech (Fig. 13.1). The specific actions of the facial muscles are not detailed here as they are amply covered by books on muscle testing.

General principles for treating the face are:

1. Gross motions are mass opening and mass closing.
2. There are two general facial areas, the eyes and forehead, and the mouth and jaw. The nose works with both general areas.
3. Facial motions are exercised in diagonal patterns.

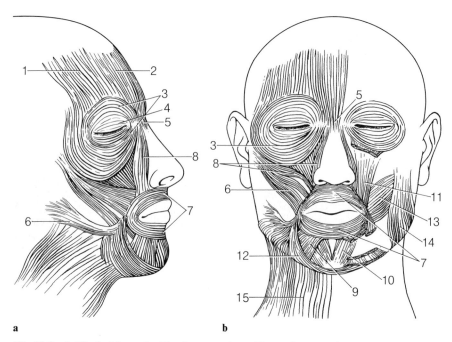

a   **b**

Fig. 13.1 a, b. The facial muscles. Numbers correlate with muscles on next pages

Fig. 13.2. A mirror can help patients control their facial movements

4. The face should be treated bilaterally; the stronger side reinforces motions on the weaker side.
5. Strong motions in other parts of the body will reinforce the facial muscles. This occurs in our everyday lives.
6. Functionally, the facial muscles must work against gravity; this must be considered when choosing a position for treatment.
7. A mirror can help patients control their facial movements (Fig. 13.2).

1. M. Epicranius (Frontalis) (Fig. 13.3)

Command. "Lift your eyebrows up, look surprised, wrinkle your forehead."

Apply *resistance* to the forehead, pushing caudally and medially. This motion works with eye opening.

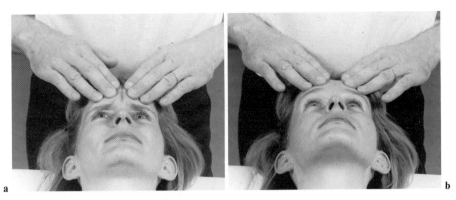

a   b

Fig. 13.3 a, b. Facilitation of m. epicranius (frontalis)

2. M. Corrugator (Fig. 13.4)

Command. "Frown, pull your eyebrows down."

Give *resistance* just above the eyebrows diagonally in a cranial and lateral direction. This motion works with eye closing.

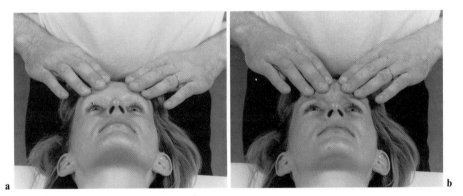

a   b

Fig. 13.4 a, b. Facilitation of m. corrugator

3. M. Orbicularis Oculi (Fig. 13.5)

Command. "Close your eyes."

Use separate exercises for the upper and lower eyelids. Give gentle diagonal *resistance* to the eyelids. Avoid putting pressure on the eyeballs.

235

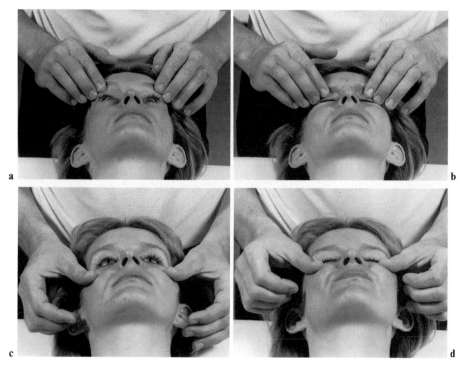

Fig. 13.5 a–d. Facilitation of m. orbicularis oculi

4. M. Levator Palpebrae Superioris (Fig. 13.6)

Command. "Open your eyes."

Give *resistance* to the upper eyelids. Resistance to eyebrow elevation will reinforce the action.

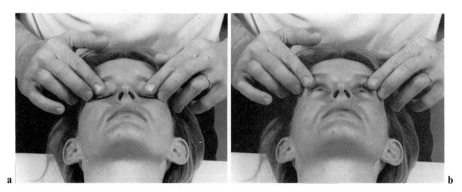

Fig. 13.6 a, b. Facilitation of m. levator palpebrae superioris

5. M. Procerus (Fig. 13.7)
Command. "Wrinkle your nose."

Apply *resistance* next to the nose diagonally down and out. This muscle works with m. corrugator.

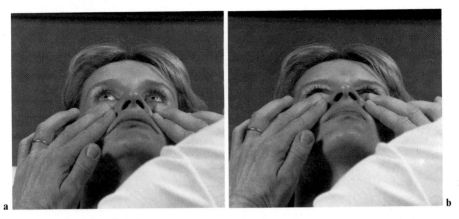

Fig. 13.7 a, b. Facilitation of m. procerus

6. M. Risorius and M. Zygomaticus Major (Fig. 13.8)
Command. "Smile."

Apply *resistance* to the corners of the mouth medially and slightly downward (caudally).

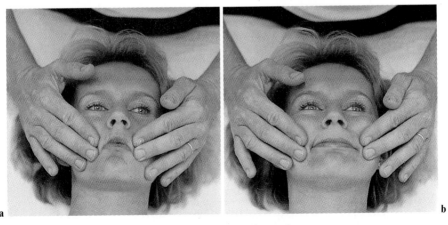

Fig. 13.8 a, b. Facilitation of m. risorius and m. zygomaticus major

7. Orbicularis Oris (Fig. 13.9)

Command. "Purse your lips, whistle, say 'prunes'."

Give *resistance* laterally and upward to the upper lip, laterally and downward to the lower lip.

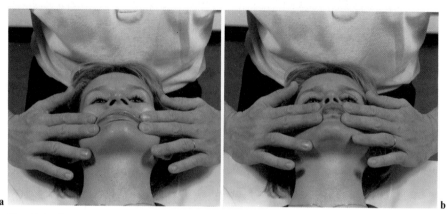

Fig. 13.9 a, b. Facilitation of m. orbicularis oris

8. M. Levator Labii Superioris (Fig. 13.10)

Command. "Lift your upper lip, show your upper teeth."

Apply *resistance* to the upper lip, downward and medially.

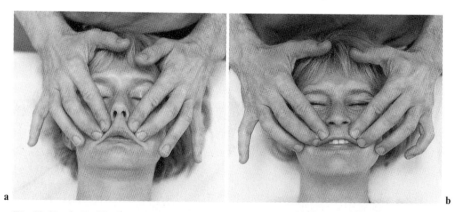

Fig. 13.10 a, b. Facilitation of m. levator labii superioris

9. M. Depressor Labii Inferioris

Command. "Push your lower lip downwards, show your lower teeth."

Apply *resistance* upward and medially to the lower lip. This muscle and the platysma work together.

238

10. M. Mentalis (Fig. 13.11)
Command. "Wrinkle your chin."
Apply *resistance* down and out at the chin.

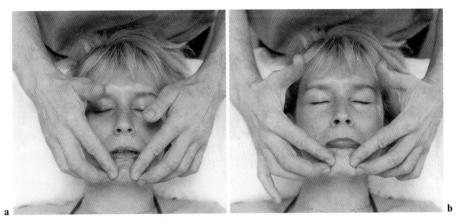

Fig. 13.11 a, b. Facilitation of m. mentalis

11. M. Levator Anguli Oris (Fig. 13.12)
Command. "Pull the corner of your mouth up, sneer."
Push down and in at the corner of the mouth.

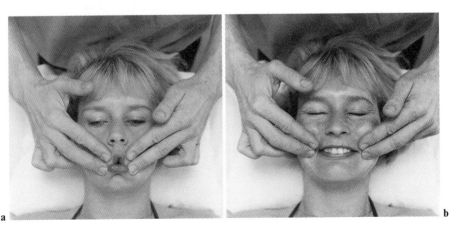

Fig. 13.12 a, b. Facilitation of m. levator anguli oris

12. M. Depressor Anguli Oris (Fig. 13.13)

Command. "Push the corners of your mouth down, look sad."
Give *resistance* upwards and medially to the corners of the mouth.

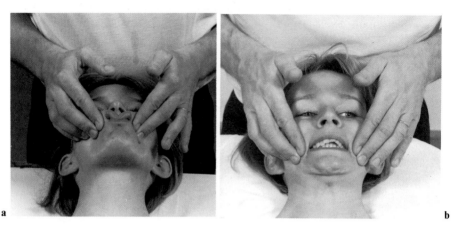

a   b

Fig. 13.13 a, b. Facilitation of m. depressor anguli oris

13. M. Buccinator (Fig. 13.14)

Command. "Suck your cheeks in, pull in against the tongue blade."
Apply *resistance* on the inner surface of the cheeks with your gloved fingers or a dampened tongue blade. The resistance can be given diagonally upward or diagonally downward as well as straight out.

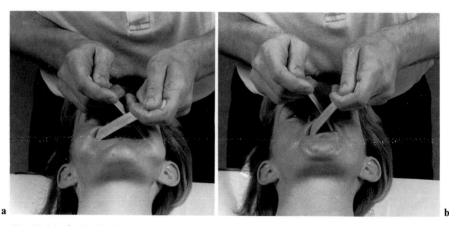

a   b

Fig. 13.14 a, b. Facilitation of m. buccinator

14. M. Masseter Temporalis (Fig. 13.15)
Command. "Close your mouth, bite."

Apply *resistance* to the lower jaw diagonally downward to the right and to the left. Resist in a straight direction if diagonal resistance disturbs the temporomandibular joint. Resistance to the neck extensor muscles reinforces active jaw closing.

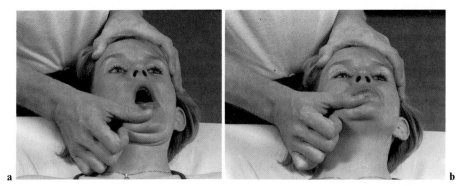

Fig. 13.15 a, b. Facilitation of m. masseter and m. temporalis

15. M. Infrahyoid and Mm. Suprahyoid (Fig. 13.16)
Command. "Open your mouth."

Give *resistance* under the chin either diagonally or in a straight direction (see 14). Resistance to the neck flexor muscles reinforces active jaw opening.

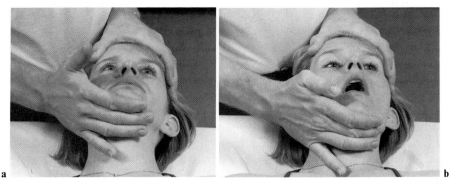

Fig. 13.16 a, b. Facilitation of mm. infrahyoidei, mm. suprahyoidei, and the platysma

16. M. Platysma (Fig. 13.17)

Command. "Pull your chin down."

Give *resistance* under the chin to prevent the mouth from opening. Resistance may be diagonal or in a straight plane as in Fig. 13.15. Resisted neck flexion reinforces this muscle.

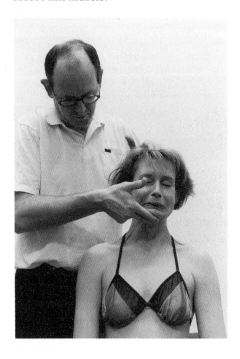

Fig. 13.17. Training of the platysma

17. Intrinsic Eye Muscles

Eye motions are reinforced by resisted head and trunk motion in the desired direction.

> *Example:* To reinforce eye motion down and to the right, resist neck flexion to the right and ask the patient to look in that direction.
>
> To reinforce lateral eye motion resist full rotation of the head to that side and tell the patient to look to that side.

Give the patient a definite target to look at with your command.

> *Example:* "Tuck your head down (to the right) and look at your right knee."

13.3 Tongue Movements

Use a tongue blade or your gloved fingers to stimulate and resist tongue movements. Wet the tongue blade to make it less irritating to the tissues. Ice the tongue to increase the stimulation. Sucking on an ice cube permits patients to stimulate tongue and mouth function on their own.

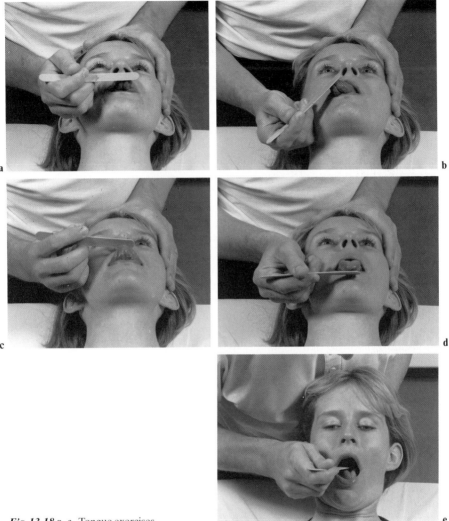

Fig. 13.18 a–e. Tongue exercises

We have illustrated the following tongue exercises:

- Sticking the tongue out straight (Fig. 13.18 a)
- Sticking the tongue out to the left and the right (Fig. 13.18 b)
- Touching the nose with the tongue (Fig. 13.18 c)
- Touching the chin with the tongue (Fig. 13.18 d)
- Rolling the tongue (Fig. 13.18 e)

Other tongue motions which should be exercised include:

- Humping the tongue (needed to push food back in the mouth in preparation for swallowing)
- Moving the tongue laterally inside the mouth
- Touching the tip of the tongue to the palate just behind the front teeth

243

13.4 Swallowing

Swallowing is a complex activity, controlled partly by voluntary action and partly by reflex activity (Kendall and McCreary 1983). Exercise can improve the action of the muscles involved in the reflex portion as well as in the voluntary portion of swallowing. Sitting, the functional eating position, is a practical position for exercising the muscles involved. Another good treatment position is prone on elbows.

Chewing is necessary to mix the food with saliva and shape it for swallowing. The tongue moves the food around within the mouth and then pushes the chewed food back to the pharynx with humping motions. To keep the food inside the mouth, patients must be able to hold their lips closed. Exercise of these facial and tongue motions is covered Sects. 13.2 and 13.3.

A hyperactive gag reflex will hinder swallowing. To help moderate this conditioned reflex, use prolonged gentle pressure on the tongue, preferably with a cold object. Start the pressure at the front of the tongue and work back toward the root. Simultaneous controlled breathing exercises will make the treatment more effective.

When the food reaches the back of the mouth and contacts the wall of the pharynx it triggers the reflex that controls the next part of the swallowing action. At the start of this phase the soft palate must elevate to close off the nasal portion of the pharynx. Facilitate this motion by stimulating the uvula with a dampened swab. You can do this on both sides, or concentrate just on the weaker side.

As the swallowing activity continues, the hyoid bone and the larynx move upward. To stimulate the muscles that elevate the larynx use quick ice and stretch. Give the stretch diagonally down to the right and then to the left. Treat hyperactivity in these muscles with prolonged icing, relaxation techniques, and controlled breathing.

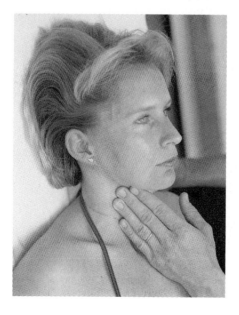

Fig. 13.19. Stimulation or relaxation of the throat

13.5 Speech Disorders

For satisfactory speech a person needs both proper motion of the face, mouth, and tongue and the ability to vary tone and control breathing. Patients who have only high vocal tones are helped by breathing exercises and ice over the laryngeal area. Patients with only low vocal tone benefit from stimulation of the laryngeal muscles with quick ice followed by stretch and resistance to the motion of laryngeal elevation. *Note:* To prevent compression of the larynx or trachea apply pressure on only one side of the throat at a time (Fig. 13.19).

Promote controlled exhalation during speech with resisted breathing exercises (Sect. 13.6). Use combination of isotonics, starting with resisted inhalation (concentric contraction) and following with resisted exhalation (eccentric contraction). During exhalation the patient recites words or counts as high as possible. Work on the patient's control of speech volume in the same way.

13.6 Breathing

Breathing problems can involve both breathing in and breathing out. Treat the sternal, costal, and diaphragmatic areas to improve inspiration. Exercise the abdominal muscles to strengthen forced exhalation.

All the procedures and techniques are useful in this area of care. Hand alignment is particularly important to guide the force in line with normal chest motion. Use stretch reflex to facilitate the initiation of inhalation. Continue with repeated stretch through range (repeated contractions) to facilitate an increase in inspiratory volume. Appropriate resistance strengthens the muscles and guides the chest motion. Preventing motion on the stronger or more mobile side, timing for emphasis, will facilitate activity on the restricted or weaker side. Combination of isotonics is useful when working on breath control. The patient should do breathing exercises in all positions. Emphasize treatment in functional positions.

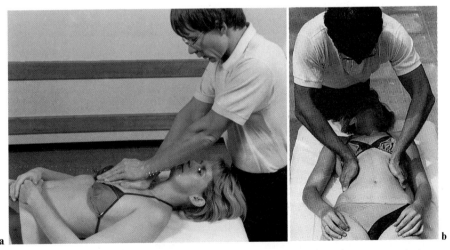

a   b

Fig. 13.20 a, b. Breathing in a supine position: *a* pressure on the sternum, *b* pressure on the lower ribs

Supine

- *Sternal:* Place both hands on the sternum and apply oblique downward pressure (caudal and dorsal, towards the sacrum) (Fig. 13.20a).
- *Ribs:* Apply pressure on the lower ribs, diagonally in a caudal and medial direction, with both hands (Fig. 13.20b). Place your hands obliquely with the fingers following the line of the ribs. Exercise the upper ribs in the same way, placing your hands on the pectoralis major muscles.

Sidelying

- *Sternal:* Use one hand on the sternum, the other on the back to stabilize and give counterpressure.
- *Ribs:* Put your hands on the area of the chest you wish to emphasize. Give the pressure diagonally in a caudal and medial direction to follow the line of the ribs. Point your fingers point in the same direction. In sidelying the supporting surface will resist the motion of the other side of the chest (Fig. 13.21).

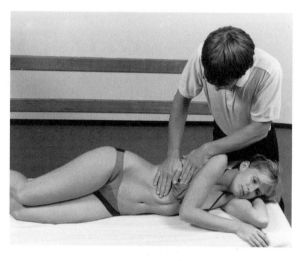

Fig. 13.21. Breathing in a sidelying position

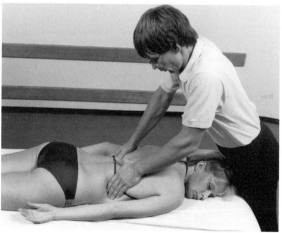

Fig. 13.22. Breathing in a prone position

Prone (Fig. 13.22)

– *Ribs:* Give pressure caudally along the line of the ribs. Place your hands on each side of the rib cage over the area to be emphasized. Your fingers follow the line of the ribs.

Prone on Elbows

– *Sternal:* Place one hand on the sternum and give pressure in a dorsal and caudal direction. Put your other hand on the spine at the same level for stabilizing pressure (Fig. 13.23).
– *Ribs:* Use the prone position hand placement and pressures.

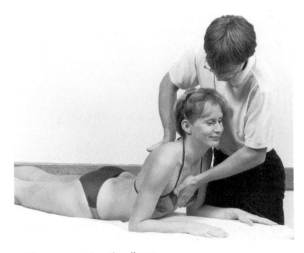

Fig. 13.23. Breathing in a prone position supported on the elbows

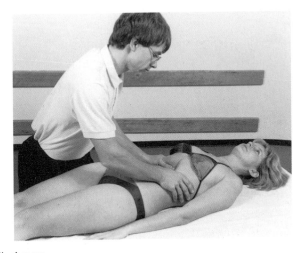

Fig. 13.24. Facilitation of the diaphragm

247

Facilitation of the Diaphragm

You can facilitate the diaphragm directly by pushing upward and laterally with the thumbs or fingers from below the rib cage (Fig. 13.24). Apply stretch and resist the downward motion of the contracting diaphragm. The patient's abdominal muscles must be relaxed for you to reach the diaphragm. To give indirect facilitation for diaphragmatic motion, place your hands over the abdomen and ask the patient to inhale while pushing up into the gentle pressure. Teach your patients to do this facilitation on their own.

Reference

Kendall FP, McCreary EK (1983) Muscles, testing and function. Williams and Wilkins, Baltimore

14 Activities of Daily Living

Mastering the activities of daily living (ADL) is an important step in the patient's progress toward independence. The previous chapters have described a range of activities for achieving this goal: mat activities (rolling, bridging, crawling, kneeling, sitting), standing, walking, head and neck exercises, facial exercises, breathing, and swallowing.

When the patient has mastered the fundamentals needed for success in ADL, time may be spent working on more advanced or difficult activities. All the skills that a patient needs for independence can be taught using the PNF treatment approach. Guidance given by grip and resistance helps the patient develop effective ways to perform these activities.

Some of the practical activities are:

– Transferring from the wheelchair into, e. g., a bed, a shower, a chair, a toilet, a car. (Fig. 14.1).
– Dressing and undressing (Fig. 14.2), washing, self-care.
– Many occupational therapy activities.

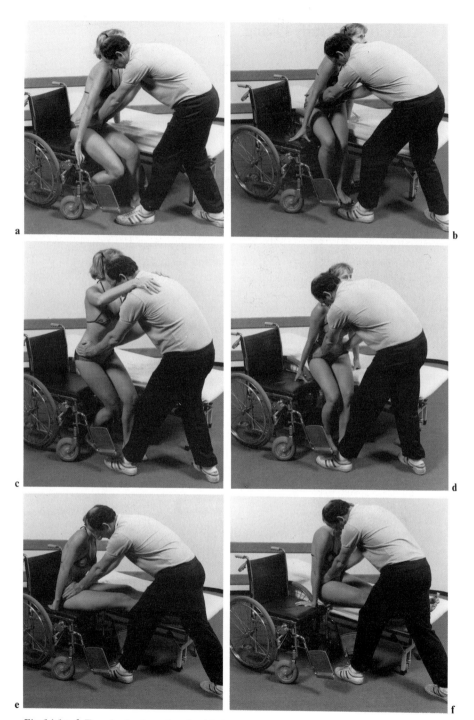

Fig. 14.1 a–f. Transferring from the wheelchair

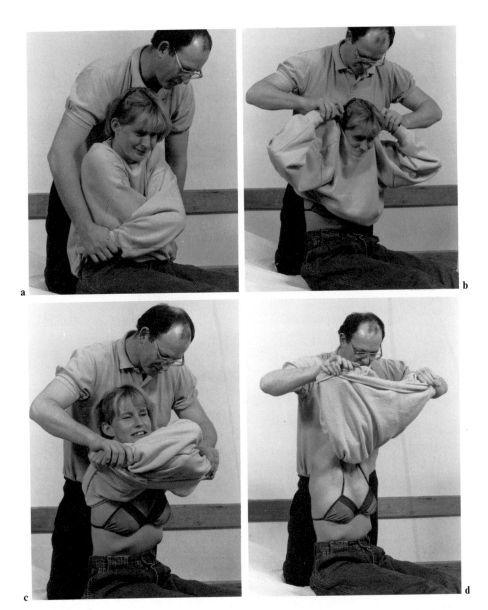

Fig. 14.2 a–d

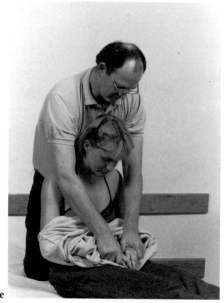

e

f

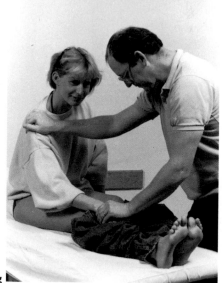

g

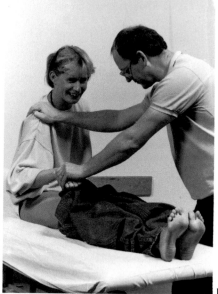

h

Fig. 14.2 e–h

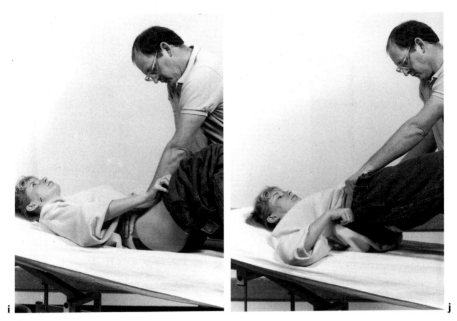

Fig. 14.2 i, j. Undressing and dressing

15 Glossary

Afterdischarge: The effect of a stimulus, such as a muscle contraction, continues after the stimulus has stopped. The greater the stimulus the longer the afterdischarge.

Approximation: The compression of a segment or extremity through the long axis. The effect is to stimulate a muscular response and improve stability.

Basic principles: The philosophy of proprioceptive neuromuscular facilitation.

Basic procedures: Tools to increase the effectiveness of treatment.

Bilateral: Of both arms or both legs.
  Bilateral asymmetrical: moving both arms/legs in opposite diagonals but in the same direction.
    Example: Right extremity, *flexion*–abduction; left extremity, *flexion*–adduction.
  Bilateral symmetrical: moving both arms/legs in the same diagonals and the same direction.
    Example: Right extremity, *flexion*–abduction; left extremity, *flexion*–abduction.
  Bilateral symmetrical reciprocal: moving both arms/legs in the same diagonals but in opposite directions.
    Example: Right extremity, *flexion*–abduction; left extremity, *extension*–adduction.
  Bilateral asymmetrical reciprocal: moving both arms/legs in opposite diagonals and in opposite directions.
    Example: Right extremity, *flexion*–adduction; left extremity, *extension*–adduction.

Chopping: Bilateral asymmetrical upper extremity extension with neck flexion to the same side to exercise trunk flexion (see Sect. 10.2.1)

Elongated state: The position in a pattern where all the muscles are under tension of elongation. Usually the starting position for the pattern.

Excitation: Activation or stimulation of muscular contractions.

Facilitation: Promoting or encouraging motor activities.

Groove/diagonal: The line of movement in which a pattern takes place. Resistance is applied in this line of movement. The therapist's arms and body line up in this groove or diagonal.

Hold: An isometric muscle contraction.

Inhibition: Suppressing or restraining muscle contractions.

Irradiation: The spread or increased force of a response that occurs when a stimulus is increased in strength or frequency. This ability is inherent to the neuromuscular system.

Lifting: Bilateral asymmetrical upper extremity flexion with neck extension to the same side to exercise trunk extension (see Sect. 10.2.2).

Lumbrical grip: A grip in which the lumbrical muscles are the prime movers. The metacarpal–phalangeal (MCP) joints flex and the proximal (PIP) and distal (DIP) interphalangeal joints remain relatively extended. Traction and rotational resistance are effectively applied with this hold.

Muscle contractions:
    Isotonic (dynamic): the intent of the patient is to produce motion.
        Concentric: shortening of the agonist produces motion.
        Eccentric: an outside force, gravity or resistance, produces the motion. The motion is restrained by the controlled lengthening of the agonist.
        Stabilizing isotonic: the intent of the patient is motion, the motion is prevented by an outside force (usually resistance).
    Isometric (static): the intent of both the patient and the therapist is that *no* motion occur.

Overflow: The expansion of a response from the stronger to the weaker parts of a pattern or from stronger patterns to weaker patterns of motion.

Pivot of action: The joint or body section in which movement takes place.

Reinforcement: The strengthening of a weaker segment by a stronger segment, which has been specially chosen for this purpose. It can work within a pattern or the reinforcement can come from another part of the body.

Repeated contractions: Eliciting the stretch reflex repeatedly from an already contracting muscle or muscles to produce stronger contractions.

Reversal: An agonistic motion followed by an antagonistic motion. This is an effective form of facilitation based on *reciprocal innervation* and *successive induction.*

Reciprocal Innervation: Excitation of the agonist is coupled with simultaneous inhibition of the antagonist. This provides a basis for coordinate movement.

Stretch: Elongation of muscular tissue
    Stretch stimulus: increased excitation in muscles in the elongated state.
    Quick stretch: a short, sharp stretch of muscles under tension. A quick stretch is required to get a stretch reflex.
    Restretch: another quick stretch to a muscle that is under tension.

Successive induction: Contraction of the antagonists is followed by an intensified excitation of the agonist. This is a basis for the reversal techniques.

Summation: The joining of subliminal stimuli resulting in excitation (contraction).

Spatial summation: Simultaneous subliminal stimuli from different parts of the body join to produce muscle contraction.

Temporal summation: The combining of stimuli that occur within a short time period to produce muscle contraction.

Technique: Resistance to muscle contraction combined with appropriate facilitory procedures to achieve specific objectives. The techniques may be combined to achieve the desired results.

Timing:

Normal timing: the course or sequence of movements that results in coordinate movement.

Timing for emphasis: changing the normal timing of movements to emphasize some component within the movement. This is especially effective when applying optimal resistance to stronger components within the movement.

Unilateral: Of only one leg or arm.